Survival Analysis

Survival analysis generally deals with analysis of data arising from clinical trials. Censoring, truncation, and missing data create analytical challenges, and the statistical methods and inference require novel and different approaches for analysis. Statistical properties, essentially asymptotic ones, of the estimators and tests are aptly handled in the counting process framework which is drawn from the larger arm of stochastic calculus. With the explosion of data generation during the past two decades, survival data has also enlarged assuming a gigantic size. Most statistical methods developed before the millennium were based on a linear approach even in the face of the complex nature of survival data.

Nonparametric nonlinear methods, commonly employed in statistical inference for analysis of survival data are best envisaged in the Machine Learning setting. This book attempts to cover all these aspects in a concise way.

Survival Analysis offers an integrated blend of statistical methods and machine learning useful in the analysis of survival data. A special feature of the offering is to give an exposure to the machine learning trends for lifetime data analysis.

Features:

- Classical survival analysis techniques for estimating relevant statistical functions and hypotheses testing

- Regression methods covering the popular Cox relative risk regression model, Aalen's additive hazards model and related diagnostics.

- Information criteria to facilitate model selection including Akaike, Bayes, and Focused

- Penalization methods covering l_1, l_2, and SCAD

- Survival trees and ensemble techniques of bagging, boosting, and random survival forests

- A brief exposure of neural networks for survival data

- R program illustration throughout the book

Prabhanjan Narayanachar Tattar has industrial experience with Dell International Services, Ford Motor Company, and British American Tobacco. The author has published several books in Statistics: A Course in Statistics with R (J. Wiley), Statistical Application Development with R and Python, and Hands-on Ensemble Learning with R. He is a recipient of the IBS(IR)-GK Shukla Young Biometrician Award (2005) and the Dr. U.S. Nair Award for Young Statistician (2007). He held Senior Research Fellowship of CSIR-UGC during PhD. In the year 2021, he has ventured into fiction writing and published three novels under the penname *S.B. Akshobhya*.

H. J. Vaman is a retired professor of statistics. He has taught the subject for over 40 years at Bangalore University and Central University of Rajasthan. He has also served as visiting faculty at Shivaji University, University of Calcutta, Indian Statistical Institute, Bangalore Centre, IIT-Mumbai, and Mangalore University. His main areas of research are sequential decision processes, survival analysis, statistical process control, and modeling in certain health-related studies.

Survival Analysis

Prabhanjan Narayanachar Tattar
H. J. Vaman

CRC Press
Taylor & Francis Group
Boca Raton London New York

CRC Press is an imprint of the
Taylor & Francis Group, an **informa** business

A CHAPMAN & HALL BOOK

First edition published 2023
by CRC Press
6000 Broken Sound Parkway NW, Suite 300, Boca Raton, FL 33487-2742

and by CRC Press
4 Park Square, Milton Park, Abingdon, Oxon, OX14 4RN

CRC Press is an imprint of Taylor & Francis Group, LLC

ISBN: 978-0-367-03037-7 (hbk)
ISBN: 978-1-032-30848-7 (pbk)
ISBN: 978-1-003-30697-9 (ebk)

DOI: 10.1201/9781003306979

Typeset in CMR10
by KnowledgeWorks Global Ltd.

TO

Our Teachers
Who Left a Lasting Imprint on our Minds

Contents

Contents

Preface

The purpose of writing this book, *Survival Analysis*, is to provide an updated account of some of the recent developments in this field. The authors intend to accomplish it from two fronts: (i) newer or extended methods in survival analysis developed over the past two decades and thereby extending the scope and applicability of existing body of methods of survival analysis contained in the currently available texts, and (ii) augmenting the traditional methods with their counterpart in machine learning. The Aalen's additive regression model, a useful alternative to the popular Cox proportional hazards model, and certain multiple variants of the additive model are dealt with in some detail in this book. Resampling methods have also been given greater coverage and, in a natural way, it lays the basis for some machine learning methods relevant to Survival Analysis discussed in the later part of the book.

The text opens with a number of clinical trials and related data sets which have been considered for analysis in several standard texts. The objective is one of familiarizing the reader with the nature of commonly encountered situations, the variables of interest, the distinctive nature of the data and the questions which one attempts to answer. Besides, these data sets serve the purpose of application of the techniques, existing and newer.

Regression methods based on pseudo-observations are treated in this work. Emphasis is also laid on the model selection topic which is addressed through a discussion of the choice of Information Criterion - Akaike, Bayes, and Focused. Alongside, we consider two modern penalization methods, LASSO, and SCAD for feature selection.

The trend and influence of Machine Learning methods are gradually becoming all pervasive in Statistics. The second part of the book treats the relevant machine learning techniques in a comprehensive manner as applicable to analysis of survival data.

Historically, Neural Network models have been developed for survival data by considering the lifetime as well as the censoring indicator as the output neurons. Recently, however, survival trees have been gaining incremental traction. These topics and the associated nuances are discussed in adequate detail. Extensions of survival trees in the form of random survival forests are especially important and this topic is addressed adequately. Use of Ensemble techniques is discussed and Illustrated for analysis of event time data.

R software has been generously used for illustrating the methods presented in the text. It is hoped that these illustrations will complement the discussion of the methodologies and initiate the reader to get a better insight and attempt their application to new data sets. Authors have assumed that the reader has a decent exposure to probability theory and R programming. Having basic knowledge of the traditional Survival Analysis methods will be an advantage. The authors will be happy to receive comments and suggestions from the readers and practitioners in the field.

Prabhanjan Narayanachar Tattar

H. J. Vaman

Author Bios

Prabhanjan Narayanachar Tattar has working industrial experience with Dell International Services, Ford Motor Company, and British American Tobacco. The author has published several books in *Statistics: A Course in Statistics with R* (J. Wiley), *Statistical Application Development with R and Python*, and Hands-on Ensemble Learning with R. He is the recipient of the IBS(IR)-GK Shukla Young Biometrician Award (2005) and the Dr. U.S. Nair Award for Young Statistician (2007). He held SRF of CSIR-UGC during his PhD. In the year 2021, he ventured into fiction writing and published three novels under the penname of S.B. Akshobhya.

H. J. Vaman is a retired professor of statistics. He taught for over 40 years at Bangaore University and Central University of Rajasthan. He has also served as visiting faculty at Shivaji University, University of Calcutta, Indian Statistical Institute, Bangalore Centre, IIT-Mumbai, and Mangalore University. His main areas of research are sequential decision processes, survival analysis, statistical process control, and modeling in certain health-related studies.

Symbol Description

T	The nonnegative lifetime RV, usually time to occurrence of an event like death, $T \geq 0$			aggregating the number of events by time t
			$Y(t)$	The at-risk process
C	The nonnegative censoring RV of an observation, $C \geq 0$		$H(t)$	The compensator, $H(t) = \int_0^t Y(s)h(s)ds$
$f(t)$	The probability density function PDF of T at time t, assumed to be an absolutely continuous function		$M(t)$	The martingale associated with N
			Γ	The states of a multi-state model (MSM), generally $\Gamma = \{1, 2, \ldots, k, 0\}$
$F(t)$	The cumulative distribution function CDF of T		k	The number of transitive states in a MSM, 0 is the absorbing
$S(t)$	The survival function of T, and thereby $S(t) = 1 - F(t)$		$X(t)$	A function which gives the state of the patient at time t
$h(t)$	The hazard rate of T			
$H(t)$	The cumulative hazard function of T, and it is $H(t) = \int_0^t h(s)ds$		$\prod$	the product-integral
			$\mathbf{P}(s,t)$	transition probability matrix of a nonhomogeneous Markov process
$g(t)$	The PDF of the censoring RV C			
$G(t)$	The CDF of the censoring RV C		$H_{jj'}(t)$	Cumulative intensity matrix associated with $\mathbf{P}(s,t)$
$\tilde{T}$	The observed value, it might be lifetime or censoring time, $\tilde{T} = \min\{T, C\}$		$K(t)$	A predictable and nonnegative weight function for setting up test statistics
			i	The indexing which varies over observations
δ	Indicator that the observed time is a lifetime, $\delta = I\left(\tilde{T} = T\right)$		j	The indexing which varies over the states of the Markov process
$L, \log L$	The likelihood function, the log-likelihood function		l	The indexing which varies over the covariates $x_1, x_2, \ldots, x_p$
$N(t)$	The counting process			

Acronyms

AFT   accelerated failure time
AIC   Akaike information criterion
ALL   acute lymphoblastic leukemia
AML   acute myeloctic leukemia
ANOVA analysis of variance
AUC   area under curve
BIC   Bayesian information criterion
CART   classfication and regression trees
CAV   cardiac allograft vasculopathy
CDF   cumulative distribution function
CGD   chronic granulotomous disease
CNN   convolutional neural networks
CT   clinical trials
CVD   cardio vascular disease
DFR   decreasing failure rate
EM   expectation-maximization
FIC   focused Information Criterion
GBM   gradient boosting method
GEE   generalized estimating equations
GVHD graft-vs-host disease
HRQoL health related quality of life
IFR   increasing failure rate
IID   independent and identically distributed
k-NN   k-nearest neighborhoods
KM   Kaplan-Meier
LASSO least absolute shrinkage and selection operator
LDA   linear discriminant analysis
LR   log-likelihood ratio
MAR   missing at random
MCAR missing completely at random
ML   machine learning
MLE   maximum likelihood estimator
MNAR missing not at random
MSM   multi-state model
NN   neural networks
OLS   ordinary least squares

PBC	primary biliary cirrhosis
PCA	principal component analysis
PDF	probability density function
PH	proportional hazard
PrBF	probability of being in response
QAL	quality adjusted lifetime
QALY	quality adjusted life years
RNN	recurrent neural networks
RV	random variable
SCAD	smoothly clipped absolute deviation
TPM	transition probability matrix
TWiST	time without symptoms of disease and toxicity

Part I

Classical Survival Analysis

Chapter 1

Lifetime Data and Concepts

1.1 Introduction

Lifetime data implies that we are dealing with studies where the data correspond to the lifetimes of units or subjects constituting a random sample from a well defined population. The term random sample implies that, generally, the mechanism generating the sample ensures independence and identical distribution. In practice, however, these requirements are nearly satisfied or that the lack of these in the strict sense does not materially affect statistical inference carried out based on such data. Since we are considering lifetimes, it is obvious that the values of the observations will be non-negative. Formally, if we denote by T the lifetime value, we have $T \geq 0$. Such observations and corresponding data are generated in clinical trials, actuarial studies, and industrial setup where the primary variable of interest is a lifetime variable. Recently, the credit industry and the telecom industry are using the methods for lifetime data to understand problems such as customer loyalty and "customer lifetime value". The examples considered in the book will be mostly from the area of clinical trials and we will loosely call the analysis as 'survival analysis'.

We mention some examples of lifetime variables:

- Time to death of a patient/volunteer in a clinical study

- Lifetime of an equipment such as electric/electronic device, photocopy machines, etc.

- Age distribution in insurance policies and the survival time till the "death" occurs.

- "Customer loyalty" of an user with a telecom company, credit card bank, megastore, etc.

- Time to relapse post recovery of a disease, say cancer.

- Time to de-addiction for unhealthy habits such as smoking, drugs, etc.
  □

Aalen, et al. (2008) [2], abbreviated often as ABG in the book, Andersen, et al. (1993), abbreviated as ABGK, [7] Fleming and Harrington (1991) [44],

DOI: 10.1201/9781003306979-1

Kalbfleisch and Prentice (2002)[60] and Lawless (2002)[69] are some of the treatises on survival analysis.

The variable of interest, lifetime, as it emerges in data has a different structure in survival analysis as compared with the usual output variable in other studies. Suppose that a pharmaceutical company is developing a drug for curing a cancer ailment. The drug helps in elimination of the tumor, and we are interested in estimating how long the patient survives after the treatment. If the drug is to be compared against a benchmark drug, or against placebo, we randomly assign a group of patients with this drug and the other group with placebo. For the moment we will assume that all the patients respond similarly to the drugs and that information on factors such as health history and other related variables are not available. Here, we cannot afford to observe/followup the patients until the event of death is observed for each of the patients. The study has to be clearly stopped either by a certain time or when a pre-decided number of events is observed, and as such the event would not be necessarily observed for all the patients. Thus the observations might be incomplete and censored by the stopping criterion.

The incomplete observations are called as *censored observations*. It turns out that the censoring is, most often, because of not knowing the time when the event would occur for the observation in future and at the best we know a minimum time when the unit had been observed as "alive". The censoring in these examples form *right censoring*. Even though the event is incompletely observed for one or more subjects, an incomplete observation does provide the information that the survival time is at least as large as the observed lifetime.

An indicator variable δ is generally used to indicate the status of the observation. The value of δ as 1 indicates that the observation has been completely observed. Its value of 0 indicates the observation has been censored, and it is then common to denote the variable by $\tilde{T}$. Observing continuously may not always be feasible. In such cases, the units in the samples are observed only at pre-determined time points or at the time of occurrence the events of interest. The patient might be lost between two follow-up times and here we refer to the censoring as *interval censoring*. In other studies, the interest might be in the survival time since the adoption of an unhealthy practice. However, many addicts might not be able to trace back the exact time of the their first use of the drugs and they might provide fuzzy answers. In such cases, the time origin of the observations is unknown and the values are subject to *truncation*, especially truncation on the left side since the initial time point is not known.

The random process associated with survival analysis can be explained through the model flexibility provided by *stochastic processes*. We will have an informal discussion of this aspect here. Consider the example of survival time T of a patient diagnosed for cardio vascular disease (CVD). Here, death due to heart attack, indicated by $\delta = 1$, will determine the value of T, and this phenomenon can be explained by a stochastic process with two states $\Gamma = \{'Alive', 'Death'\}$. Any time before the event $\delta = 1$, the patient is in the *Alive* state, and if the person dies due to heart failure, the shift occurs from

Alive state to the *Death* state. The event *Death* is an absorbing state since a transition out of the state is not possible. Since time is a continuous variable, we have a continuous stochastic process with two states. At any given time t, let $\zeta(t), t \geq 0$, denote the state of the patient. In the formal stochastic calculus terminology, T will act as a *random stopping time*.

In the examples considered thus far, the event occurs because of a specific cause. For example, in the cancer treatment study, the event is death due to cancer. If the death is even caused by any other reason, say accident, we say that the observation is censored. However, in quite some studies, the patients are treated for multiple ailments and the event 'causes' is more than one. For example, the patient might be suffering from multiple types of cancer. In this case, we have multiple causes which are competing for the event and such studies are known as *competing risks* and their modeling as *competing risks modeling*. This class of models have been studied in detail, see David and Moeschberger (1978)[34].

The competing risks phenomenon too can be explained by using the stochastic process framework. Here, we have multiple causes of risks and the occurrence of any one of it pre-empts the occurrence of other causes. Clearly, the state space can be then extended to $\Gamma = \{'Alive','Death\ 1','Death\ 2',\ldots,'Death\ k'\}$, where k is the different number of ways the event can occur. As earlier, at any given time, the state of the patient is given by $\zeta(t)$. Note that the k death states are all absorbing states, and we do not have any possible transitions once any of the state is entered.

The competing risks model can be viewed as a special case of the generic *multi-state model*. Let $\Gamma = \{1, 2, \ldots, k\}$ be the k states of a multi-state model. Here, a patient can transit from any state $i \in \Gamma$ to any other state $j \in \Gamma$. If $k = 2$ and the state 2 is an absorbing state, we have the generic time-to-event type of study. If only state 1 denotes the alive state and the rest of states denote some absorbing states, we have the competing risks model. Before delving into more details, we will have a look at some of the popular datasets from the literature.

1.2   Survival Datasets

Datasets are important and we need them in the study of clinical trials and survival analysis. The datasets described in this section will be used in two ways: (i) to illustrate various concepts of survival analysis and the related methodology, and (ii) to solve, more importantly, real-life problems for which purpose the data had been collected in the first place. The description provided here will setup the context of the problem and the variables that are present in the data.

TABLE 1.1:   Lifetimes
of 463 Sheeps

Age at Death	Frequency
0–2	121
2–3	7
3–4	8
4–5	7
5–6	18
6–7	28
7–8	29
8–9	42
9–10	47
0–11	66
11–15	90

1.2.1   Lifetimes of Sheep

Anil P Gore, Sai Paranjape, and Madhav Kulkarni had collected and published over hundred datasets, and they are available in the R package **gpk**, see also https://cran.r-project.org/web/packages/gpk/index.html. The dataset **Sheeplife** contains the lifetimes of 463, see Table 1.1. Here, we do not have the exact lifetime values of the 463 sheep, and we only know the time interval in which the sheep had died. Here, we do not have any censored observations. A parametric estimate of the mean lifetime might suffice to summarize the lifetime of a sheep.

As simple as the data appears in Table 1.1, we can see two major challenges. First, as mentioned earlier, though we do not have the censored observations, the lifetime values are incompletely recorded and at best we know the lower bound and upper bound on the actual lifetime values. The incomplete information can be handled using the *Expectation-Maximization algorithm*, or simply the EM algorithm. The problem will be specifically addressed in Chapter 2.

It can also be seen in Table 1.1 that the number of deaths is very high in the intial period of 0–2 years and this likely explains the infant mortality. The period of 2–7 years sees a sharp decline in the mortality rate, and it is also almost constant in this period. The mortality rate increases from age 7 onwards.

1.2.2   Mayo Clinic Primary Biliary Cirrhosis Study

Mayo Clinic has conducted many clinical trials and a popular dataset widely used is the primary biliary cirrhosis (PBC) of the liver conducted during the period 1974–84. The primary biliary cirrhosis problem is known in recent times as primary biliary cholangitis. It is the autoimmune disease of liver. This problem arises because of a slow and progressive destruction of the small bile ducts of the liver leading to a build-up of bile and other toxins in the liver.

The purpose of the clinical trial was to find whether the drug D-penicillamine improves the lifetime of the patients treated for the PBC problem. A total of $n = 424$ patients satisfied the eligibility criteria. Six patients were immediately lost to follow-up post the diagnosis of the ailment. A total of 106 patients did not choose to participate in the trial and they consented only to provide basic measurements and follow-up for the survival event. The remaining 312 patients agreed to participate in the randomized clinical trial. The main variables measured, available for 17 more variables, for the patients are described next:

- Age: the age of the patient

- Sex: the gender of the patient

- Treatment: indicator of whether the patient received the D-penicillamine drug, placebo, or opted out of the trial and gave only basic measurements and survival follow-up

- Platelet: the platelet count

- Stage: the histologic stage of disease obtained by biopsy

- Measurements such as serum albumin,alkaline phosphotase, presence of ascites, aspartate aminotransferase, serum bilirubin, serum cholesterol, urine copper, edema, hepato, blood vessel malformations in the skin

- Time: the number of days between registration and the earlier of death, transplantation, or study analysis in July, 1986

- Status: status at endpoint, 0/1/2 for censored, transplant, dead

We need to analyze how effective has been the treatment of D-penicillaime drug in improving the lifetime of the patients inflicted by the primary biliary cirrhosis problem. Apart from identifying the effectiveness of the drug, we are also interested in understanding the impact of the systolic and related physiological measurements on the patients lifetime. A brief summary of the variables is provided in Table 1.2.

The Role of Covariates. In the introductory discussion, we did not mention anything about the covariates, or additional information. Clinical trials will most often have a lot of patient health history and it is also important to incorporate them in the analysis. Covariates play a very important role

Survival Analysis

TABLE 1.2: The Variables Summary of Primary Biliary Cirrhosis

Variable Name	Variable Type	Missing Observations	Range/Levels
ID	Name		
Time	Continuous	0	(41,4795)
Status	Indicator	0	0, 1, 2
Treatment	Indicator	0	1,2
Age	Continuous	0	(26.28,78.44)
Sex	Indicator	0	Male, Female
Presence of ascites	Indicator	106	0, 1
Hepato (Enlarged liver)	Indicator	106	0,1
Spiders	Indicator	106	0,1
Edema	Continuous	0	(0,1)
Serum bilirubin	Continuous	0	(0.3,28)
Serum cholesterol	Continuous	134	(120,1775)
Serum albumin	Continuous	0	(1.96,4.64)
Urine copper	Continuous	108	(4,588)
Alkaline phosphotase	Continuous	106	(289,13862.4)
Aspartate aminotransferase	Continuous	106	(26.35,457.25)
Triglycerides	Continuous	136	(33,598)
Platelet count	Continuous	11	(62,721)
Blood clotting time	Continuous	2	(9,18)
Histologic stage of disease	Indicator	6	1, 2, 3, 4

and influence the value of T and accordingly the statistical methodology of regression analysis. We would like to especially see which of the covariates are significant in explaining the survival time of the patients. The vitality of covariates in survival analysis is even more pronounced as some of them might even be time varying and we call them as *time-dependent covariates.*

1.2.3 Chronic Granulotomous Disease Study

This study consists of data from a placebo controlled trial of gamma interferon in chronic granulotomous disease (CGD). It consists of data on the time to serious infections observed through end of study for each patient. This study is an example where the event can occur more than once to the patient. After the infection is removed or treated with, we observe the patient up to the time when the infection recurs and again give the treatment. The goal is to understand if the gamma interferon treatment extends the time to recurrence of the infection.

A summary of the variables is provided in Table 1.3. The dataset is available in Appendix D of Fleming and Harrington (1991)[44].

TABLE 1.3: The Chronic Granulotomous Disease Data Summary

Variable Name	Variable Type	Range/Levels
Identification	Name	
Enrolling center	Categorical	13 different levels
Date of randomization	Date	1989-06-07 to 1989-12-29
Treatment	Indicator	"placebo" or "gamma interferon"
Sex	Indictor	Male, Female
Age	Continuous	(1,44)
Height (in cm at study entry)	Continuous	(76.3,189)
Weight (in kg at study entry)	Continuous	(10.4,101.5)
Pattern of inheritance	Indicator	X-linked, autosomal
Use of steroids (at study entry)	Indicator	0,1
Use of prophylactic antibiotics (at study entry)	Indicator	0,1
Hospital categorization	Indicator	US:NIH, US:other, Europe:Amsterdam, Europe:other
Time start	Continuous	(0,373)
Time stop	Continuous	(4,439)
Enumerate	Integer	1, 2, 3, 4
Status	Indicator	0,1

1.2.4 Bone Marrow Transplant Data for Leukemia

Leukemia is the cancer of body's blood-forming tissues and generally it includes the bone marrow and the lymphatic system. A standard treatment is the bone marrow transplant. Post transplantation, the patient is closely observed until the patient can be declared as "cured", and this period is called the remission time. The bone marrow transplant is a complex operation and the recovery following the operation is also critical. A transplantation is marked as failure if the patient's leukemia relapses or if death is observed within the remission time. On the other hand, the success of transplantation is guided by the recovery process which in turn depends on two intermediate events: (i) development of acute graft-vs-host disease (GVHD) within the first 100 days and (ii) recovery of platelet count to a level marked as more than 400 million per liter. The recovery course also depends on risk factors such as the time of tranplantation, the leukemia stage, age and gender of the patient, the time from diagnosis to transplantation, etc. Following the surgery, the recovery also depends on variables such as occurrence of acute/chronic

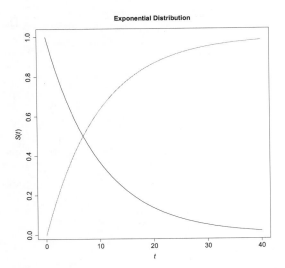

FIGURE 1.1: Survival and Cumulative Distribution Functions for Exponential Distribution*

graft-vs-host disease (GVHD), return of platelet count to normal levels, return of granulocytes to normal levels, or development of infections. See Section 1.3 of Klein and Moeschberger (2003)[64].

Figure 1.1 of Klein and Moeshberger gives a clear description of the bone marrow transplant problem. Following the transplant, the patient might see a platlet recovery or an acute GVHD. It is also possible that the transplant might succeedingly see either relapse of the problem or even death. A patient that sees platelet recovery following the transplant might then suffer from GVHD, or the disease might relapse, or even die. Similarly, a patient observed in GVHD post the transplant might see a recovery of the platelet count, or the relapse or death state.

The study is also an example of multicenter trial where the patients are readied for the transplantation through a radiation-free conditioning regimen. The patients belonged to three different disease groups: acute lymphoblastic leukemia (ALL) and acute myeloctic leukemia (AML) at high and low levels. The patients were treated at four different hospitals, see the row Hospital in Table 1.4. The count and distribution of other categorical variables can be found in the same table. Unlike the earlier three studies, the terminal event can either be relapse or death, and occurrence of one precludes that of other. Hence, as indicated in the previous section, we have an example of *competing risks model* here. Since we have lot of categorical variables, one might be interested in the response of the patients across the different disease groups over the various categorical data. Comparison of survival curves is in order and we will be using this study throughout the book.

---

* All R codes, color diagrams, and data files will be available at www.crc____.com

TABLE 1.4: Bone Marrow Transplant for Leukemia Dataset Summary

Variable Name	Variable Type	Range/Levels	Levels Frequency
Group Disease	Indicator	1(ALL), 2(AML-Low), 3(AML-High)	(38,54,45)
Time to Death	Continuous	(1,2640)	
Disease-free Survival Time	Continuous	(1,2640)	
Time To Acute GVHD	Continuous	(1,2640)	
Time To Chronic GVHD	Continuous	(1,2640)	
Time To Chronic GVHD	Continuous	(1,1298)	
Death	Indicator	0,1	(56,81)
Relapse	Indicator	0,1	(95,42)
Disease Free Survival	Indicator	0,1	(54,83)
Acute GVHD	Indicator	0,1	(111,26)
Chronic GVHD	Indicator	0,1	(76,61)
Platelet Recovery	Indicator	0,1	(17,120)
Age of patient	Continuous	(7,52)	
Donor age	Continuous	(2,56)	
Patient Sex	Indicator	0,1	(57,80)
Donor Sex	Indicator	0,1	(49,88)
Patient CMV status	Indicator	0,1	(69,68)
Donor CMV status	Indicator	0,1	(79,58)
Waiting time to transplant	Continuous	(24,2616)	
FAB	Indicator	0,1	(92,45)
Hospital	Indicator	1,2,3,4	(76,17,23,21)
Use of MTX	Indicator	0,1	(97,40)

In the next subsection, we will face with the scenario of complete multi-state model.

1.2.5 Heart Transplant Monitoring Data

Sharples, et al. (2003)[105] collected data related to analyze the diagnostic accuracy of coronary angiopathy and risk factors for post-heart-transplant cardiac allograft vasculopathy. The heart transplant monitoring dataset consists of a series of approximately yearly angiographic examinations of heart transplant recipients. Here, the state at each time is a grade of cardiac allograft vasculopathy (CAV), a deterioration of the arterial walls. The dataset has 622 unique patients and 2846 measurements among them.

TABLE 1.5:   Cardiac Allograft Vasculopathy Study

Variable Name	Variable Type	Range/Levels	Levels Frequency	Missing
Patient Number	Nominal			
Age	Continuous	6.304,74.332		
Years after transplant	Continuous	0,19.46		
Donor Age	Continuous	0,61		
Sex	Indicator	0,1	535,87	
Primary diagnosis	Indicator	CVCM, Hyper, IDC, IHT, Other, Restr	88, 7, 1283, 1413, 10, 15	30
Acute rejection episodes (cumulative)	Continuous	0-12		
State	Indicator	1, 2, 3, 4	2039, 351, 205, 251	
First observation	Indicator	0,1	2224, 622	
Maximum observed state	Indicator	1, 2, 3, 4	1985, 405, 232, 251	

A summary of the variables of this dataset is provided in Table 1.5. As mentioned earlier, the state at each time is a grade of CAV with the higher number reflecting a deterioration of the arterial walls. The information available on prognostic variables includes the age of the patient, the number of years since following the transplant, donor age, the patient gender, and primary diagnosis for the reason of transplant. For each further change of state, we record here the cumulative number of acute rejection episodes, the state at the examination, maximum observed state up to the time. Multistate models will be used to analyze the CAV data.

1.2.6   Netherlands Cancer Institute Seventy Gene Signature

In all the studies considered earlier, the number of observations were much larger than the number of covariates or explanatory variables. With data collection gathering a considerable momentum during the past two decades, increasing exponentially, a certain technical problem props up in a number of studies. We have covariates closer to a significant proportion of the number of observations, and in some settings nearly equal to it. The covariates count may exceed the number of observations in other studies. Sometimes, this problem is called as *curse of dimensionality*. Note that we are not yet referring to the case where more covariates are generated, say transformations by square term, or some other polynomial order. Here, the variable of interest is metastasis-free time among lymph node positive breast cancer patients.

In this study, 144 patients were followed up for the metastasis-free time. The occurrence of metastasis or death is the follow-up event and a few observations are censored. Five regular type of covariates information is collected on the patients. The age of the patient at the time of cancer treatment, grade of the tumor at one of three levels, estrogen receptor status, number of affected lymph nodes, and diameter of the tumor are recorded for the patients. In addition, we have gene expression measurements of 70 prognostic genes. With just 144 patients, difficulties arise in carrying out analysis and inference. This dataset will be addressed in Chapter 7.

1.3 Basic Survival Analysis Concepts

The main variable of interest in survival analysis is the time with respect to an origin to the type of event that's being observed to occur for an experimental unit. As a consequence of the structure of the problem, the variable will be a non-negative random variable and we will denote it by T throughout the text. By definition, we have $T \geq 0$. The associated probability space and related mathematical requirement will be introduced in the later chapters. In this brief section, we will consider three core concepts survival function, hazard rate, and cumulative hazard function.

1.3.1 Survival Function

Here, we will assume that T is a non-negative absolutely continuous random variable. The scenario when T is a vector will be considered very briefly in a later development. Suppose then that F is the cumulative distribution function of T, denoted by $T \sim F$, that is, $0 \leq F(t) \leq 1, t \geq 0$, otherwise, and let f denote the corresponding probability density function. It is also important to note that T in certain cases will be neither continuous, nor discrete. For instance, a clinical trial following an operation, some patients might not make it through the operation round itself, and thereby $T = 0$ for a considerable number of patients. In such cases T might be considered as a *mixture random variable*.

The probability of the patient living at time t is often referred as the *survival probability* and this probability is denoted by $S(t)$, that is,

$$S(t) = P(T > t), \tag{1.1}$$

which is also related to the cumulative distribution function $F(t)$ by

$$S(t) = P(T > t) = 1 - P(T \leq t) = 1 - F(t).$$

Survival Analysis

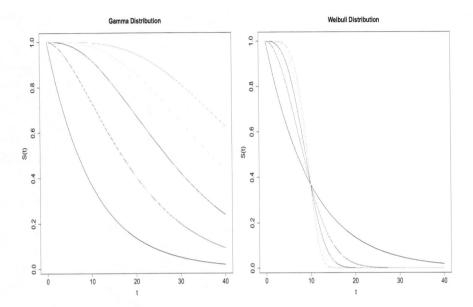

FIGURE 1.2: Gamma and Weibull Survival Functions

The survival probability $S(t)$ is widely known as the *survival function*, and is complement of $F(t)$. Note that $S(0) = 1$.

Figure 1.1 shows the survival function as well as the cumulative distribution function of the famous *exponential distribution*. The exponential distribution is characterized by a single parameter, known as *failure rate*, and it has many properties that make it attractive in the applications of survival analysis. The exponential distribution is still the default choice when it comes in applications of survival analysis as a first approximation. Apart from the applications in survival analysis and queueing theory, it is useful in reliability analysis and software failure analysis.

The strength of the exponential distribution is in its simplicity, and that itself becomes a bane in other setups. Since the distribution is governed by a single parameter, it will not be able to explain the data when there is more variability across the observations. Generalization of the exponential distribution is thus important and unequivocally required. Gamma and Weibull distributions are important alternative contenders for the exponential distribution and Figure 1.2 shows a plot of the survival function for these two distributions with different choices of parameters. Details of the parameters and exact form follow in the next chapter.

The survival function gives useful probabilities. In the next part of the section, we will dwell on two other important functions.

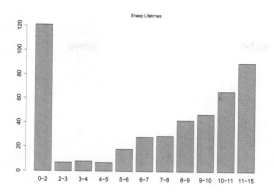

FIGURE 1.3: Sheep Lifetime Distribution

1.3.2 Hazard Rate and Cumulative Hazard Function

In Subsection 1.2.1, we have come across the lifetimes of 463 sheep and the data was binned in intervals of years. Figure 1.3 depicts the bar diagram of the count of lifetimes observed in the binned intervals. Note that the number of sheep dying in the first two years is very high, and the next bins of 2–3, 3–4, and 4–5 years shows fewer deaths. After the initial high death rate, there is a sharp decline, a few years of stability, and then gradual progression in the death rates. A measure of death counts as provided by either the survival function of the cumulative distribution might not be adequate. Thus, for lifetime data, we have a special requirement of the failure rate.

The *instantaneous failure rate*, or *hazard rate*, of a non-negative random variable T with survival function $S(t)$ is defined by

$$
\begin{aligned}
h(t) &= \lim_{\delta \downarrow 0} \frac{P(t < T \le t + \delta)}{\delta}, \\
&= \lim_{\delta \downarrow 0} \frac{S(t) - S(t + \delta)}{\delta}, \\
&= \lim_{\delta \downarrow 0} \frac{F(t + \delta) - F(t)}{\delta}.
\end{aligned}
\tag{1.2}
$$

The formula for hazard rate $h(t)$ given by Equation 1.2 can be expressed in different and equivalent ways. It can be obtained as a limiting function of the cumulative distribution function as well as the survival function.

Figure 1.4 gives the plot of hazard function $h(t)$ against t for different distributions and different parameters. For the exponential distribution, the hazard rate is constant across time. For Weibull and gamma distribution, the

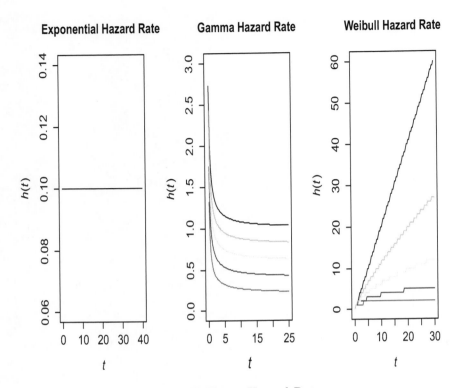

FIGURE 1.4: Hazard Rates

parameter values can be changed to obtain different curves for the hazard rate function.

The *cumulative hazard function*, denoted by $H(t)$, is defined by

$$H(t) = \int_0^t h(s)ds. \tag{1.3}$$

The mathematical relationship between hazard rate, cumulative hazard function, and the survival function will be explored in the next chapter. The cumulative hazard function for different distributions are given in Figure 1.5.

Parametric statistical inference of the survival models will be considered in the later part of the book.

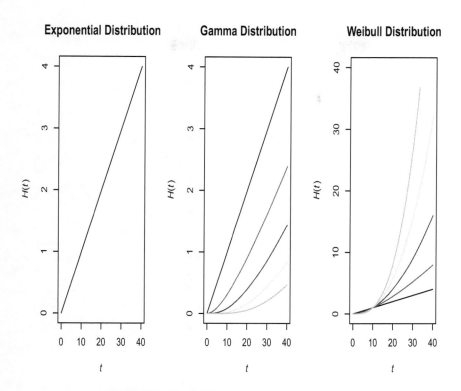

FIGURE 1.5: Cumulative Hazard Function

1.4 Statistical Inference for Survival Data

The survival function, hazard rate, and cumulative hazard functions are useful when the parameters of the underlying model are completely known. When the parameters are unknown, the data is collected and the parameters are estimated. The complexities of truncation, censoring, and missing data compel different strategies of statistical inference for survival data as against the traditional parametric inference.

With time to event data and indicators of whether the observations are complete or not, we have the useful nonparametric methods of inference concerning the survival curves. For instance, we can estimate the survival function or the cumulative hazard function using nonparametric methods. The survival function is estimated using the famous *Kaplan-Meier estimator*. Kaplan-Meier is often listed as one of the twenty important methods of the previous century. It provides an estimate of $S(t)$, denoted by $\hat{S}(t)$ and it is often reliable

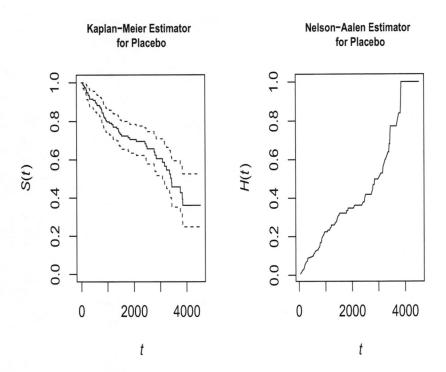

FIGURE 1.6: Kaplan-Meier and Nelson-Aalen Estimators

until the time point of the last complete observations. It is also known as the *product-limit estimator*. The justification of the name follows later.

For the PBC data, see Section 1.2.2, we can fit Kaplan-Meier estimator of the survival function. The corresponding cumulative hazard function $H(t)$ is estimated by using the famous *Nelson-Aalen estimator*. The plot of Kaplan-Meier and Nelson-Aalen estimators for the PBC data is given in Figure 1.6.

In the PBC problem, we have two types of treatment, and we would like to test the hypotheses whether the survival functions for the patients under the two treatment arms are equal, or whether they are different. Figure 1.7 shows the result, that is, it gives the plot of the Kaplan-Meier survival functions for both the treatment arms.

The two survival curves in Figure 1.7 are clearly different from one another. However, we would like to know whether the differences are significant or not. A log-rank test can be then applied to test whether the two survival curves are equal to one another. The next R code snippet carries out the logrank test.

Comparing Survival Curves

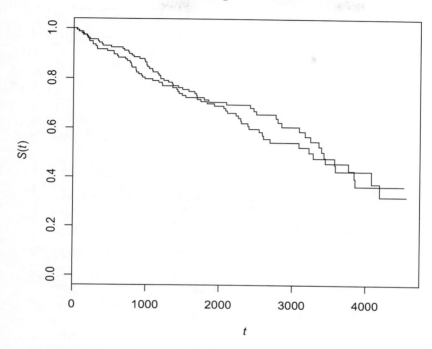

FIGURE 1.7: Kaplan-Meier Estimators for Two Treatment Arms

```
> survdiff(Surv(time,status==2)~trt,data=pbc)
Call:
survdiff(formula = Surv(time, status == 2) ~ trt, data = pbc)

n=312, 106 observations deleted due to missingness.

          N Observed Expected (O-E)^2/E (O-E)^2/V
trt=1 158       65     63.2    0.0502     0.102
trt=2 154       60     61.8    0.0513     0.102

 Chisq= 0.1  on 1 degrees of freedom, p= 0.75
```

The *p*-value is seen at 0.75 which means that there is not enough evidence in the data to conclude that the survival curves for the treatment arms are distinct.

Recollect that in Table 1.2 we have given a host of variables that can be very vital in explaining the survival times under the treatment arms. Regression models are not straightforward to build for survival data. For instance, the least-squares criterion will produce inconsistent estimators of the coefficients of the covariates. It is in the specification of the covariates influence on the lifetimes that the notion of hazard rate comes very effective. In fact, regression modeling of lifetimes in terms of covariates was proving to be a very difficult task until the the advent of the path breaking work of Sir D R Cox (1972) [28] proposing a semi-parametric model approach through the hazard rate as follows:

$$h(t|\beta, x) = h_0(t) \exp \left\{ \beta' x \right\}. \tag{1.4}$$

Here, $h_0(t)$ is the baseline hazard rate. The model specified in Equation 1.4 is known as the *proportional hazards model* or the *relative risk model*. The following R code snippet shows the fitted Cox proportional hazards model.

```
> summary(pbc_PH)
Call:
coxph(formula = Surv(time, status == 2) ~ ., data = pbc[, -1])

  n= 276, number of events= 111
   (142 observations deleted due to missingness)

              coef  exp(coef)   se(coef)        z Pr(>|z|)
trt     -1.242e-01  8.832e-01  2.147e-01   -0.579  0.56290
age      2.890e-02  1.029e+00  1.164e-02    2.482  0.01305 *
sexf    -3.656e-01  6.938e-01  3.113e-01   -1.174  0.24022
ascites  8.833e-02  1.092e+00  3.872e-01    0.228  0.81955
hepato   2.552e-02  1.026e+00  2.510e-01    0.102  0.91900
spiders  1.012e-01  1.107e+00  2.435e-01    0.416  0.67760
edema    1.011e+00  2.749e+00  3.941e-01    2.566  0.01029 *
bili     8.001e-02  1.083e+00  2.550e-02    3.138  0.00170 **
chol     4.918e-04  1.000e+00  4.442e-04    1.107  0.26829
albumin -7.408e-01  4.767e-01  3.078e-01   -2.407  0.01608 *
copper   2.490e-03  1.002e+00  1.170e-03    2.128  0.03337 *
alk.phos 1.048e-06  1.000e+00  3.969e-05    0.026  0.97893
ast      4.070e-03  1.004e+00  1.958e-03    2.078  0.03767 *
trig    -9.758e-04  9.990e-01  1.333e-03   -0.732  0.46414
platelet 9.019e-04  1.001e+00  1.184e-03    0.762  0.44629
protime  2.324e-01  1.262e+00  1.061e-01    2.190  0.02850 *
stage    4.545e-01  1.575e+00  1.754e-01    2.591  0.00958 **
---
Signif. codes:  0 '***' 0.001 '**' 0.01 '*' 0.05 '.' 0.1 ' ' 1
```

The fitted model reinforces the earlier observation that the treatment arm is insignificant. The significant variables are age of the patient, edema, serum bilirubin, serum albumin, urine copper, aspartate aminotransferase, blood clotting time, and the histologic stage of disease. In the summary output, the more number of asterisks (*s) indicates higher statistical significance.

In practice, it is advisable to remove the insignificant variables from the fitted model, and after removing them, we get the following output:

```
coxph(formula = Surv(time, status == 2) ~ age + edema + bili +
    albumin + copper + ast + protime + stage, data = pbc_na[,
    -1])
```

```
              coef exp(coef)  se(coef)      z       p
age       0.031384  1.031881  0.010204   3.08  0.0021
edema     0.821795  2.274580  0.347146   2.37  0.0179
bili      0.085121  1.088849  0.019335   4.40  1.1e-05
albumin  -0.718595  0.487436  0.272449  -2.64  0.0084
copper    0.002854  1.002858  0.000983   2.90  0.0037
ast       0.004377  1.004386  0.001807   2.42  0.0154
protime   0.227517  1.255479  0.101373   2.24  0.0248
stage     0.432794  1.541558  0.145631   2.97  0.0030

Likelihood ratio test=164  on 8 df, p=0
n= 276, number of events= 111
```

Thus, only significant variables are now retained in the model.

1.5 Machine Learning Inception

Machine learning has evolved as an important and alternative paradigm to statistical methods in analysis of data. The paradigm derives strong theoretical justificaion based on the concepts of *empirical risk minimization*, the *probably approximately correct*, and the *Vapnik-Chervonenkis dimension*. Neural networks, support vector machine, fuzzy logic, and decision trees are some of the methods in the so-called *supervised learning* arm of machine learning. In supervised learning we have a target variable, also equivalent to the regressand variable in the traditional linear models, and the purpose is to understand such a variable in terms of the explanatory or independent variable. Unsupervised learning as an arm of machine learning is not relevant in our study. The decision trees have been developed with great exposition in the Breiman school of machine learning.

The main attraction of machine learning methods is that it packs powerful nonlinear methods without requiring the user to specify the mathematical form of nonlinearity. In fact, the objective of most machine learning methods is to enable machines, such as computers and other computing devices, to take an appropriate decision. If deployed in the true spirit, it does not require human intervention after rigorous testing. A widely held perception is that machine learning is a kind of black box method that is not tractable mainly on account of its implementation. One of the goals of the book is demystify such impressions and allow the reader to clearly understand the underpinnings of machine learning, and more especially in the context of survival data.

In this brief section, we will outline in very informal terms how a decision tree is created. In the general setup, we have n observations and p variables. To begin with, all the observations are held in the root node. Since the decision tree is a machine learning method, it is empirically driven. The generic tree algorithm takes one variable at a time. Consider a variable, say X_1, and determine all unique values in the data. Let d_1 denote the number of unique values, and we will denote the unique values by $x_{11}, x_{12}, \ldots, x_{1d_1}$. For each of the d_1 values, the observations are partitioned in two regions: the right side and the left side by the following logic. If the X_1 value of the observation is greater than $x_{1k}, k = 1, \ldots, d_1$, the observation is moved to the right partition, else to the left. For a regression problem, we calculate the residual sum of squares in each part and then add them. The unique value of X_1 that leads to the least residual sum of squares is chosen as the split point for the variable. The split point for X_1 can be denoted by x_{1s}. We note that x_{1s} is one of the values from $x_{11}, x_{12}, \ldots, x_{1d_1}$. The exercise is repeated for other variables $X_2, \ldots, X_p$ and we obtain $x_{2s}, \ldots, x_{ps}$. Finally, we select that variable which has the least residual sum of squares among $x_{1s}, \ldots, x_{ps}$. The overall data is then partitioned according to this variable. The partitions are also called as nodes of the trees. The exercise is repeated recursively for each of the partitions, equivalently the tree is grown, until no further improvement is possible. The nodes when the tree can not be split further are called as terminal nodes. For classification problem, the criteria is by either of Gini index, Bayes error, or the cross-entropy measure. For survival data, the criteria is determined by the log-rank statistic.

An implementation of the survival tree in the R software is provided by statistician Terry Therneau. An application of the function `rpart` from the package of the same name `rpart` on the PBC dataset leads to Figure 1.8. In the survival tree, the first split criterion is to move to the right side if the serum bilirubin value is less than 2.25 mg/dl, else we move the left side. We proceed similarly until we end up at one of the terminal node which is called as the leaf node.

The survival data here is internally scaled by the `rpart` package so that the predicted rate in the root node (the first split point) is unit value. Relative to this value, the values in the terminal node reflect the hazard rate.

At each split point, the variable that has the minimum log-rank value will be selected and stored in the background. After the complete tree is created, the values across the tree are then added up to obtain the *variable importance*. The next R code snippet gives the variable importance of the survival tree constructed for the PBC dataset. The serum bilirubin is the best variable with maximum importance value followed by the protime, and so forth and aspartate aminotransferase is the least significant variable.

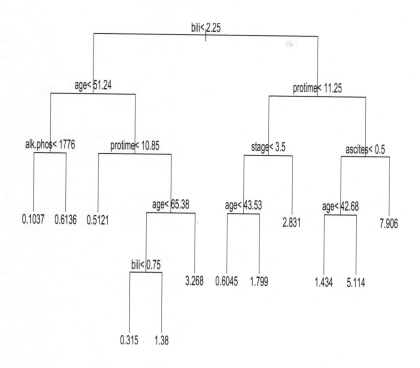

FIGURE 1.8: PBC Survival Tree

```
> pbc_stree$variable.importance
     bili   protime        age  albumin     edema      stage
138.0078   70.8673   54.5482  32.2399   25.5762   15.2313
  ascites alk.phos platelet       sex    copper        ast
 14.0942   13.4409   10.0179   2.4528    2.1149    1.6919
```

The details of the survival tree will follow in the second part of the book.

1.6 Roadmap

The nature of survival data has been elaborated through different datasets in Section 1.2. The section introduced various concepts such as interval censoring, inferring the functions of interest such as cumulative hazard function and survival function, comparison of survival curves, impact of covariates on the hazard rates, etc. Sections 1.3–1.5 give a brief peek in addressing the problems. We will now outline the approach implemented in the book.

The book is in two major parts, and part I covers the discussion from Chapters 1–7.

Chapter 2 will introduce the survival concepts in depth. The probability models useful for the domain will also be dealt here. Univariate and bivariate lifetime models will be discussed here. Extensions of the univariate lifetime distributions is also considered. Censoring and the impact on the likelihood function under noninformative censoring is detailed. The role of EM algorithm is briefly dealt with an application to the sheep lifetimes. The counting process approach will give a brief preview on its importance with an R program. The initial theory of multi-state models will be given an exposition.

Chapter 3 develops nonparametric estimation of survival functions. Estimation of the cumulative hazard function and the survival function respectively through the famous Nelson-Aalen and Kaplan-Meier estimators will be given in succession. These techniques hold good under the assumption of noninformative censoring. Mean and median estimation is carried out based on Kaplan-Meier estimator. Estimation of the hazard function is as difficult a problem as estimation of density function. The hazard function estimation is based on the Nelson-Aalen estimator and smoothing techniques. The chapter will close with a detailed illustration of the building of the multi-state models. The transition probability matrix of the associated nonhomogeneous Markov process will be estimated by using the Aalen-Johansen estimator.

Chapter 4 develops statistical tests for comparison of the survival times in the clinical trial studies. The graphical display of the difference between two survival curves as seen in Figure 1.7 might or might not be significant. The output of the R program following that figure clearly shows that the difference is insignificant. In this chapter, a diverse family of survival tests will be developed to carry out appropriate hypothesis testing problems.

The all important regression methods will be developed in Chapters 5 and 6. The chapter will begin with linear regression models extended for survival data, and is followed by the Cox proportional hazards model, or the relative risk model. Extensions to time-varying covariates will be included too. Residual analysis following regression modeling is vital and the parallels of linear regression models are provided here. The assumption validation follows parametric regression models. Aalens' additive risk model is a complementary topic. Pseudo-observations are emerging as an important alternative to the regular survival regression models, and we demonstrate these models while augmenting the weighted regression models. Since all covariates will not turn out to be significant, it is important to chunk out the insignificant variables. The topic of *model selection* forms the core of Chapter 7. In addition to the regular Akaike Information Criterion (AIC) and Bayesian Information Criterion (BIC) model selection criteria, the recent development in the works of Nils Lid Hjort's *Focused Information Criterion* will be illustrated in the chapter. An alternative to information criteria is provided by the penalization techniques of ℓ_1, ℓ_2 and Singularly Clipped Absolute Deviation (SCAD) penalties.

The second part, Part II will deal with the modern machine learning methods as applicable in Survival Analysis. Chapter 8 begins with a brief review of the different methods used for construction of survival tree. The logic of construction of a survival tree, as typically shown in Figure 1.8, will be discussed here. The concept of variable importance will be clearly illustrated and we will perform quick diagnostics on the survival tree. An extension of the survival tree to the multi-state model will also be introduced here. Chapter 9 includes the extension of the survival tree to ensemble methods of bagging and random forest. The chapter will develop the boosting method for survival data. Chapter 10 addresses neural network methods for survival data.

Chapter 2

Core Concepts

2.1 Introduction

Chapter 1 presented us with various classes of problems. The nature of the data and the objectives will require the analyst to choose an appropriate technique for further analysis. The core concepts of the lifetime data including hazard rate, cumulative hazard function, and survival function were introduced in Section 1.3. In the next section, we will deal with the notions in greater detail.

The fundamental concepts of hazard rate and survival function have been introduced in Chapter 1, *Lifetime Data and Concepts*. For the commonly used lifetime distributions exponential, gamma and Weibull the survival function plots are shown in Figures 1.1 and 1.2. Similarly, Figure 1.4 *Hazard Rates* gives the plot of hazard rate for the same distributions. In the next section, we will introduce the models with more mathematical details for these concepts in the context of the widely used lifetime distributions. The important properties and applications of these will be demonstrated through practical situations. The univariate distribution will be extended to the bivariate case in the next section. The generalizations of classical lifetime distributions such as exponential and Weibull, can be achieved through *resilience* and *tilt* parameters and this will be explored in Section 2.3. These modified lifetime distributions will be referred to as *generalized lifetime distributions*.

The important idea of *censoring* is introduced in Section 2.4 and we will briefly see the impact on the likelihood function and the need for simplifying assumptions such as *non-informative censoring*. Because of the complexities of the likelihood function due to censoring and truncation, the statistical inference runs into issues and we need a different framework to handle this. The counting process approach to lifetime data is debuted in Section 2.6.

The observations until this point of the chapter experience the event of interest only once, or only a single type of event is relevant. Competing risks, or exposure to different causes of death, is a study where the observation might succumb to one of the different risks. In certain studies, it is possible that the patient is not observed in the same health state until the end of the study. For instance, in prolonged diseases such as AIDS and cancer, the patient might be observed in a perfect health state following an operation. However,

DOI: 10.1201/9781003306979-2

in between the operation time and the perfect health state, the patient would have spent varied number of weeks in poor health. Further, death as the event of interest might follow the perfect health state following to a relapse of disease or deteriorating health states. Thus, we need to consider the different health states and such modeling can be carried out using *multi-state models* as briefly discussed in Section 2.7.

2.2   Lifetime Distributions

Nelson (1982)[88], Elandt-Johnson and Johnson (1980)[42], Kalbfliesch and Prentice (1980-2002)[60], and Cox and Oakes (1984)[30] express the need for concepts specifically required for the use of analysis of lifetime data. The time to event of interest might be for more than one component, and thereby we need generalizations too. We will begin with univariate lifetime events and then traverse to the multivariate setup.

2.2.1   Univariate Lifetime Distributions

We continue to denote the lifetime random variable by T and assume it to be continuous throughout the text. Let $f(t)$ denote the probability density function (PDF) of T, while $F(t)$ and $S(t)$ will respectively denote the cumulative distribution function (CDF) and survival function. Recollect the definitions from Equations (1.1)–(1.3):

$$
\begin{aligned}
S(t) &= P(T > t) = 1 - F(t) = \int_t^\infty f(s)ds, \\
h(t) &= \lim_{\delta \downarrow 0} \frac{S(t) - S(t+\delta)}{\delta} = \frac{f(t)}{S(t)} = -\frac{d}{dt} log(S(t)), \\
H(t) &= \int_0^t h(s)ds.
\end{aligned}
$$

The above interrelations between the three quantities $S(t)$, $h(t)$, and $H(t)$ are useful and given the form of one of them, the other can be derived. It can also be proved that $S(t) = \exp\{-H(t)\}$. We will consider the simplest lifetime distribution, viz. the exponential distribution. Suppose T denotes the lifetime of an individual, or the time to event of interest, and the probability distribution follows exponential distribution with failure rate λ, that is,

$$
f(t|\lambda) = \begin{cases} \frac{1}{\lambda} \exp\left(-\frac{t}{\lambda}\right), & t \geq 0, \lambda > 0, \\ 0, & \texttt{otherwise.} \end{cases} \tag{2.1}
$$

The simple integration of the $f(s|\lambda)$ over the interval $[0, t)$ gives the CDF of T as

$$F(t) = 1 - \exp\left(-\frac{t}{\lambda}\right),$$

and it is further straightforward to see that the survival function for the distribution is

$$S(t) = \exp\left(-\frac{t}{\lambda}\right).$$

The hazard rate of the exponential distribution is seen to be

$$h(t) = \frac{f(t)}{S(t)} = \frac{\frac{1}{\lambda}e^{-t/\lambda}}{e^{-t/\lambda}} = \frac{1}{\lambda}.$$

The hazard rate of the exponential distribution is interesting since it is independent of the time component. This was also seen in the plot of 'Exponential Hazard Rate' remains flat/constant in Figure 1.4. The constant rate is also a *characterization* of the exponential distribution, that is, the property is unique to it. Table 19.1 of Chapter 19 of Johnson, et al.[59] (1995) gives three equivalent properties unique to the exponential distribution, and the most famous characterization of this distribution is the *memoryless property*, that is, for all $t, s \geq 0$, we have $P(T > t + s | T > s) = P(T > t)$. The reader can verify this assertion. Note that the memorylessness does not imply that for $t > s$ $P(T > t | T > s) = P(T > t)$, a subtle difference.

The cumulative hazard function of this lifetime distribution is

$$H(t) = \frac{t}{\lambda}, t > 0.$$

The mean and variance of the exponential distribution 2.1 are respectively given by

$$E(T) = \lambda, \tag{2.2}$$
$$Var(T) = \lambda^2. \tag{2.3}$$

Table 2.1 gives additional details of the exponential distribution, like mode, percentile, etc. The same set of the functions are also listed in the table for Weibull, gamma, and lognormal distributions.

A distribution F is said to be an *increasing failure rate*, IFR, distribution if $h(t)$ increases in time t, that is, for $t > t'$, $h(t) > h(t'), \forall t, t'$. Similarly, we say that F has *decreasing failure rate* (DFR) distribution if the hazard rate is decreasing in time, $t > t'$, $h(t) < h(t'), \forall t, t'$. The hazard rate of exponential distribution is constant λ, and hence it is neither IFR nor DFR. We will next look at the hazard rates of Weibul and gamma distributions.

The PDF and hazard rate of the gamma distribution are given by:

$$f(t, \lambda, \beta) = \frac{1}{\Gamma(\beta)} \frac{t^{\beta-1}}{\lambda^\beta} \exp\left(-\frac{t}{\lambda}\right),$$

$$h(t, \lambda, \beta) = \frac{1}{\Gamma(\beta)\lambda^\beta} \frac{t^{\beta-1} \exp(-t/\lambda)}{1 - \Gamma(t/\lambda, \beta)}.$$

TABLE 2.1: Lifetime Distributions

Distribution	PDF	CDF	$S(t)$	$h(t)$
Exponential	$\frac{1}{\lambda}\exp\left(-\frac{t}{\lambda}\right)$	$1-\exp\left(-\frac{t}{\lambda}\right)$	$\exp\left(-\frac{t}{\lambda}\right)$	$\frac{1}{\lambda}$
Weibull	$\{\frac{\beta}{\lambda^\beta}\}t^{\beta-1}\exp\{-\left(\frac{t}{\lambda}\right)^\beta\}$	$1-\exp\{-\left(\frac{t}{\lambda}\right)^\beta\}$	$\exp\{-\left(\frac{t}{\lambda}\right)^\beta\}$	$\frac{\beta}{\lambda}\left(\frac{t}{\lambda}\right)^{\beta-1}$
Gamma	$\frac{1}{\Gamma(\beta)}\frac{t^{\beta-1}}{\lambda^\beta}\exp\left(-\frac{t}{\lambda}\right)$	$\Gamma(t/\lambda,\beta)$	$1-\Gamma(t/\lambda,\beta)$	$\frac{1}{\Gamma(\beta)\lambda^\beta}\frac{t^{\beta-1}\exp(-t/\lambda)}{1-\Gamma(t/\lambda,\beta)}$
Lognormal	$\frac{1}{\sqrt{2\pi}t\sigma}\exp\left\{-\frac{(\log(t)-\mu)^2}{2\sigma^2}\right\}$	$\Phi\left\{\frac{\log(t)-\mu}{\sigma}\right\}$	$1-\Phi\left\{\frac{\log(t)-\mu}{\sigma}\right\}$	$\frac{1}{t\sigma}\frac{\phi(\log(t)-\mu)/\sigma)}{1-\Phi(\log(t)-\mu)/\sigma)}$

We have not defined the gamma function $\Gamma(.)$. The complete gamma integral
is

$$\Gamma(\beta) = \int_0^\infty t^\beta e^{-t}dt.$$

The incomplete gamma integral $\Gamma(s,\beta)$ is

$$\Gamma(s,\beta) = \frac{1}{\Gamma(\beta)}\int_0^s u^{\beta-1}e^{-u}du.$$

Now, consider the inverse of the hazard rate, see page 604 of Ross (2010), to
obtain the following:

$$
\begin{aligned}
h^{-1}(t) &= \frac{S(t)}{f(t)} \\
&= \int_t^\infty e^{-(u-t)/\lambda}\left(\frac{u}{t}\right)^{\beta-1}du \\
&\text{with change of variables s = u-t} \\
&= \int_0^\infty e^{-s/\lambda}\left(1+\frac{s}{t}\right)^{\beta-1}ds.
\end{aligned}
$$

The function h^{-1} is monotone in the term $(1+s/t)^{\beta-1}$, and the direction
of the increase/decrease depends on the value of β. For a value of β in the
unit interval, the term $(1+s/t)^{\beta-1}$ will be increasing in time, and hence the
hazard will be decreasing or DFR. Similarly, for the value of β exceeding 1,

the term will decrease in time, and consequently the gamma hazard rate will be increasing or IFR. In the next short R program will illustrate the term. A brief digression on the choice of software.

The prominent open source software R is useful in the day-to-day analysis for a statistician. The software has well developed techniques for handling analysis of survival data. A comprehensive list of tools is available at the link `https://cran.r-project.org/web/views/Survival.html`. The authors assume the reader is familiar with the R basics, and if that is not the case, kindly refer Dalgaard (2008)[33] or Tattar (2017)[111].

The R program begins with a formulation of the mathematical function $e^{-s/\lambda} (1 + s/t)^{\beta-1}$ in the `hr`, and the function is then evaluated over the range of $(0, \infty)$ and the value is saved in `hrt`. For fixed values of λ and β, the function `ghr` returns the hazard rate at a five time t. We run the function for different values of β and fixed λ at multiple time points and check for the monoticity.

```
> # Understanding Gamma Hazard Rate (ghr function)
> ghr <- function(t,lambda,beta){
+    hr <- function(x) exp(-x/lambda)*(1+x/t)^(beta-1)
+    hrt <- 1/integrate(hr,0,Inf)$value
+    return(hrt)
+ }
> # For fixed lambda and beta < 1
> ghr(t=10,lambda=0.1,beta=0.25)
[1] 10.07427
> ghr(t=15,lambda=0.1,beta=0.25)
[1] 10.04967
> ghr(t=10,lambda=0.1,beta=0.25) < ghr(t=15,lambda=0.1,beta=0.25)
[1] FALSE
> ghr(t=10,lambda=0.1,beta=5)
[1] 9.604079
> ghr(t=15,lambda=0.1,beta=5)
[1] 9.735135
> ghr(t=10,lambda=0.1,beta=5) > ghr(t=15,lambda=0.1,beta=5)
[1] FALSE
```

As expected, the value of `ghr` for $\beta < 1$ is decreasing in time, and for $\beta > 1$, it is increasing. The hazard rate for the Weibull distribution is $(\beta/\lambda)(t/\lambda)^{\beta-1}$, which is decreasing in time for $\beta < 1$ and increasing when $\beta > 1$.
Recollect the plots of hazard and survival functions for the exponential, gamma, and Weibull distribution given in Figures 1.1, 1.2, and 1.4. It is easy to obtain such plots using R software for chosen values of the parameters. The next R session will help in generating the plots.

We first consider the plot given in Figure 1.1. A lot of plots related to the standard distributions can be easily produced using the built-in functions in the software. The general information about distributions and the software details can be obtained at `https://cran.r-project.org/web/views/Distributions.html`. To be brief here, there are four letters which are important concerning the R programming. Those four letters are p, d, q, and r, which respectively stand for the PDF, CDF (or CMF), quantiles, and random (simulation) to be used in conjuction with the standard distributions. Run `?Distributions` in your R session to get more details.

```
> pdf("../Output/Exponential_Distribution.pdf",
+ height=10,width=10)
> curve(1-pexp(x,1/10),0,40,ylab="S(t)",xlab="t")
> curve(pexp(x,1/10),0,40,add=TRUE,col="red")
> title(main="Exponential Distribution")
> dev.off()
null device
          1
```

The function `pexp` gives the cdf of exponential distribution, and the step `1-pexp` the survival function. We have invoked the graphical device using the `pdf` function, which first creates an empty file of the specified name and continues to write all plots of new page until `dev.off()` stops and saves the file.

We have discussed hazard rate and survival function without indulging in data. Let us consider a practical dataset in the Airlines industry. Frank Proschan is a renowned authority on reliability analysis and he had popularized a dataset related to the failure times of airconditioning equipment of ten Boeing 720 airplanes. We carry out visualization of the airconditioner data.

Example 1 *Visualization of Boeing Aircondition Data. The data is taken from Cox and Snell (1984)[32], and it is available in the file* Boeing.csv. *For each aeroplane, the successive failure times of the airconditioning equipment is captured in the number of hours of functioning. The number of failures for the 10 planes varies from 9 to 30, and the cumulative number of hours of functioning is in the range of 1297 to 2422.*

We will plot the histogram of the failure times for each aeroplane and superimpose an exponential distribution with failure rate equal to the inverse of the mean of the failure times. The rational for using the mean will soon be soon clarified in the chapter. The following R program accomplishes the task.

```
> # Fitting Exponential Distribution to Boeing's data
> ACD <- read.csv("../Data/Boeing.csv",header=TRUE)
> pdf("../Output/Boeing_Aircondition_Histograms.pdf")
> par(mfrow = c(1,2))
> for(i in 1:ncol(ACD)) {
+   hist(as.numeric(na.omit(ACD[,i])),
```

```
+          main=paste("Histogram of Aircondition",i),
+          xlab=paste("Aircondition",i),freq = FALSE)
+     curve(dexp(x,1/mean(as.numeric(na.omit(ACD[,i])))),
+          col = 5, lty = 2, lwd = 2, add = TRUE)
+ }
> dev.off()
null device
          1
```

The data is imported from the CSV file (comma separated variables) and stored in the R data frame object ACD. *Since the number of failure times varies across the planes, the empty cells for a particular aeroplane is read as* NA, *meaning not available, and hence when we are required to plot the histogram for the airconditioner failure times of an aeroplane, we use to command* na.omit *and convert the column into a numeric vector with* as.numeric, *we obtain the histogram using the* hist *function. The graphical parameters* main, xlab, *and* freq *allow us to respectively specify the title, x-label, and the (empirical) probability density. The* curve *function with the option of* add = TRUE *allows to superimpose the exponential distribution with rate inverse of the mean on the histogram produced in the preceding step. The reader should carefully go through the ten histograms before reading further. The ten histograms are saved in the* Boeing_Aircondition_Histograms.pdf *file, and we will evaluate the output related to the second and tenth aeroplanes.*

For the first aeroplane, see Figure 2.1, the histogram fitted with exponential distribution appears a good fit, while the tenth aeroplane visual display suggests the tail might be heavier and we need to look at alternatives to the exponential distribution.

□

Given a random sample of size n, how do we fit a lifetime distribution? Under the assumption of a random sample are the identical and independent observations $T_1, T_2, \ldots, T_n$ drawn from a continuous distribution $f(t, \theta), t \geq 0, \theta \in \Theta$, the goal is to estimate the unknown values of the parameter θ and Θ is the parameter space. The parameter θ might be scalar, or a vector. For instance, in the case of observations being drawn from an exponential distribution, we have $\theta = \lambda$, whereas if the observations are from Weibull distribution, the parameter vector is $\theta = (\lambda, \beta)$. We need general techniques of obtaining estimates of the parameters, and two popular choices are *moment estimator* and *maximum likelihood estimator*. The reader might consult Casella and Berger (2002)[22] or Rohatgi and Saleh (2015)[97]. We will denote an estimator of θ based on the random sample by $\hat{\theta} =: \hat{\theta}(t_1, \ldots, t_n)$. For a random sample, the *likelihood function* is given by

$$L(\theta, t_1, t_2, \ldots, t_n) = \prod_{i=1}^{n} f(t_i, \theta),$$

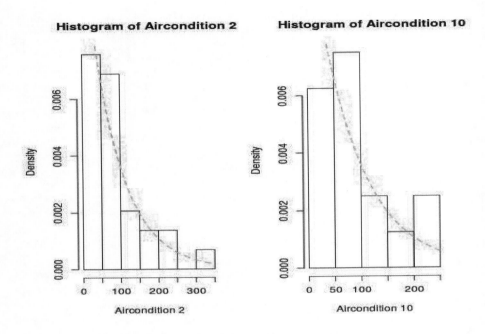

FIGURE 2.1: Histogram for Airconditioners 2 and 10

and the *maximum likelihood estimate* is that value of θ for which $L(\theta, t_1, \ldots, t_n)$ is a maximum:

$$\hat{\theta}_{mle} = \hat{\theta} = \sup_{\theta \in \Theta} L(\theta, t_1, \ldots, t_n).$$

The *likelihood principle* is among the two core principles of statistical inference, and the other important principle is that of *sufficiency*. As such the definition does not help in finding the (maximum likelihood estimators) MLEs and we need the calculus techniques in deriving an MLE. In simpler cases, the exponential distribution for example, the MLE exists in closed form expression. An elegant expression eludes in most scenarios and we then need numerical optimization methods.

Statistical inference of survival data poses certain mathematical problems in establishing the asymptotic properties. The problems that arise because of censoring can be seen in Section 2.4, and the missing data can be handled to an extent using the EM algorithm, a simple exposition of the algorithm is provided in Section 2.5. There are limitations to application of EM algorithm and hence we need to resort to nonparametric and semiparametric methods. The core technique to statistical inference of survival data will be carried out using the *counting process* approach and it will be introduced in Section 2.6.

The derivative of the log-likelihood function is known as the *score function* and it is $\delta \log L(\theta, t_1, \ldots, t_n)/\delta\theta$. Setting the derivative to zero gives the likelihood equation whose root(s) yields MLE. In the single parameter case, search for MLE can be achieved through various methods such as simple iteration, bisection, Newton-Raphson, secant, Illinois etc. Whenever a life distribution has multiple parameters, descent methods driven by Newton-Raphson method are helpful. The reader can refer Chapter 9 of Monahan (2011)[83] for more details, or Chapter 7 of Longhai Li available at `https://math.usask.ca/~longhai/teaching/2018/stat812/compstat-LLI-1.pdf`.

The MLE $\hat{\theta}_{mle}$ has several useful properties, such as consistency, asymptotic normality as well as asymptotic optimality. In formal mathematical terms, the properties translate to the following:

- As $n \to \infty$, $\hat{\theta}_{mle} \to \theta$, in probability.

- As $n \to \infty$, $\hat{\theta}_{mle} \to N\left(\theta, (nI(\theta))^{-1}\right)$, in distribution, where

$$I(\theta) = E\left(\frac{d^2}{d\theta^2} \log f(t|\theta)\right).$$

In the case of exponential distribution with pdf $f(t|\lambda) = \exp(-t/\lambda)/\lambda$, the likelihood function for a sample of size n is $L(\lambda, t_1, \ldots, t_n) \propto \exp(-\sum_{i=1}^n t_i/\lambda)$ and taking the log of the function and diffeentiating with respect to λ and equating the score function to zero, the root of the likelihood function gives the MLE in the closed form expression as $\hat{\lambda} = \bar{t} = \sum_{i=1}^n t_i/n$. This is the justification for choosing the parameter in Example 1.

Now, MLE is one method of obtaining an estimate of θ which we are denoting by $\hat{\theta}_{mle}$. Suppose there is another way of obtaining an estimate of the unknown parameter, say $\hat{\theta}_{M2}$. For instance, the ML estimator of λ is the sample mean for the exponential distribution, and we may use another method as the median of sample lifetimes, that is, $\hat{\theta}_{M2} = \mathtt{median}\{t_1, t_2, \ldots, t_n\}$. How do we then compare $\hat{\theta}_{mle}$ against $\hat{\theta}_{M2}$? The conventional statistical method is to dive into the sampling distribution of the estimators and evaluate the asymptotic relative efficiency. However, we will not dabble in this process for the primary reason that very soon censoring will kick-in and we will resort to nonparametric and semiparametric methods. Comparisons need to be performed all the same and a convenient way out is to calculate *Akaike Information Criterion*, AIC, and it is defined by

$$AIC = -2\log L(\hat{\theta}, \mathtt{data}) + 2k,$$

where k is the number of fitted parameters.

We will next fit exponential distributions to the various aeroplanes air conditioner data.

Example 2 *Visualization of Boeing Aircondition Data.*

We will use the `fitdistr` *function from the* `MASS` *package to fit an expo-*
nential distribution using the MLE technique. Since we already know that the
MLE of the failure rate λ *is nothing but the reciprocal of the mean, we will*
validate the result too. The loglikelihood function is given in the fitted R object
and we use it to calculate the AIC.

```
 library(MASS)
> t1 <- as.numeric(na.omit(ACD[,1]))
> t1_exp <- fitdistr(t1,densfun = "exponential")
> t1_exp
      rate
  0.010449796
 (0.002178933)
> -2*t1_exp$loglik + 2 # AIC
[1] 257.814
> 1/mean(t1)
[1] 0.0104498
```

Now, suppose we estimate the failure rate based on median, that is, the
reciprocal of the median as an estimate. How does the AIC then look like?
The loglikelihood function based on n observations is $-n \log \theta - \frac{\sum_{i=1}^{n} t_i}{\theta}$ *and*
using the `length` *function for the number of observations in a numeric vector,*
we quickly obtain the AIC based on median:

```
> mean(t1); median(t1)
[1] 95.69565
[1] 57
> 1/median(t1)
[1] 0.01754386
> -2*(length(t1)*log(1/median(t1))-sum(t1)/median(t1))+2
[1] 265.2084
```

It is clear that the MLE technique leads to the least AIC under the as-
sumption that the exponential distribution is the true governing probability
distribution. The reader should experiment with an estimate of the failure rate
larger than mean of the lifetimes. □

2.2.2 Multivariate Lifetime Distributions

The event of interest might be more than one in many studies. For example,
following a cataract operation, the time it takes for a pair of eyes to blurr
might be at different times for the two eyes. The lifetimes of identical twins
raised under near identical scenarios might be different. The pair of lamps
on the headlights of a car will have varying lifetimes while the time to bust

for the four tires will be different, but correlated. We will specify parametric distributions for the bivariate lifetime variables here, and an extension of the exponential distribution will be taken up. It is to be noted that there are innumerable possibilities of extending the univariate exponential distribution to the bivariate case, and it also applies to Weibull and gamma lifetime distribution.

The simplest form of survival function of bivariate exponential is given in Marshal and Olkin (1966)[75], https://apps.dtic.mil/sti/pdfs/AD0634335.pdf. Let $\mathbf{T} = (T_1, T_2)$ denote the bivariate survival times with T_j denoting the j-th component. The Marshall and Olkin bivariate exponential survival function is given by

$$
S(\mathbf{T}) = S(T_1 > t_1, T_2 > t_2)
$$
$$
= \begin{cases} \exp\left(-\frac{t_1}{\lambda_1} - \frac{t_2}{\lambda_2} - \frac{\max(t_1, t_2)}{\lambda_{12}}\right), & t_1, t_2 \geq 0, \lambda_1, \lambda_2, \lambda_{12} > 0, \\ 0, & \texttt{otherwise.} \end{cases} \quad (2.4)
$$

The Marshal-Olkin bivariate exponential distribution can be derived as a *fatal shock* model. Here, $\lambda_1, \lambda_2, \lambda_{12}$ are the parameters of the bivariate distribution. We will plot multiple survival functions corresponding to chosen values of the parameters against time t_1 and t_2 to get a holistic view. The next R program experiments with various combinations of values of $\lambda_1, \lambda_2, \lambda_{12}$ to get a peek in the behavior of $S(T_1 > t_1, T_2 > t_2)$. The time epochs for t_1 and t_2 are initialized using the seq function over the interval [0,10]. All possible combinations of t_1 and t_2 are generated and computing the expression $S(t_1, t_2)$ through the line of code exp(-t1/lambda1-t2/lambda2-pmax(t1,t2)/lambda12), we use scatterplot3d graphical command function and plot the survival function St1t2 against the various values of t_1 and t_2.

```
> # Bivariate Exponential Distribution
> t1 <- rep(seq(0,10,0.5),100)
> t2 <- rep(seq(0,10,0.5),each=100)
> pdf("../Output/Bivariate_Exponential_Distributions.pdf",
+    height=10,width=10)
> par(mfrow=c(2,2))
> lambda1 <- 1/0.4; lambda2 <- 1/0.2; lambda12 <- 1e5
> St1t2 <- exp(-t1/lambda1-t2/lambda2-pmax(t1,t2)/lambda12)
> P13D <- scatterplot3d(t1,t2,St1t2,highlight.3d=TRUE,
+    xlim=c(0,10),ylim=c(0,10),zlim=c(0,1),xlab=expression(t[1]),
+    ylab=expression(t[2]),zlab="S(t1,t2)",
+    main = expression(paste("A 3-d plot for ",
+    "Bivariate Exponential Distribution")))
> lambda_legend <- expression(paste(lambda[1],"=1/0.4"),
+                       paste(lambda[2],"=1/0.2"),
+                       paste(lambda[12],"=1/1e5"))
```

```
> legend(5,5,1,P13D$xyz.convert(18,0,12),legend = lambda_legend)
>
> lambda1 <- 1/0.4; lambda2 <- 1/0.2; lambda12 <- 1/2
> St1t2 <- exp(-t1/lambda1-t2/lambda2-pmax(t1,t2)/lambda12)
> P23D <- scatterplot3d(t1,t2,St1t2,highlight.3d=TRUE,
+   xlim=c(0,10),ylim=c(0,10),zlim=c(0,1),xlab=expression(t[1]),
+   ylab=expression(t[2]),zlab="S(t1,t2)",
+   main = expression(paste("A 3-d plot for ",
+   "Bivariate Exponential Distribution")))
> lambda_legend <- expression(paste(lambda[1],"=1/0.4"),
+                             paste(lambda[2],"=1/0.2"),
+                             paste(lambda[12],"=1/2"))
> legend(5,5,1,P23D$xyz.convert(18,0,12),legend = lambda_legend)
>
> lambda1 <- 1/0.4; lambda2 <- 1/0.025; lambda12 <- 1/0.075
> St1t2 <- exp(-t1/lambda1-t2/lambda2-pmax(t1,t2)/lambda12)
> P33D <- scatterplot3d(t1,t2,St1t2,highlight.3d=TRUE,
+   xlim=c(0,10),ylim=c(0,10),zlim=c(0,1),xlab=expression(t[1]),
+   ylab=expression(t[2]),zlab="S(t1,t2)",
+   main = expression(paste("A 3-d plot for ",
+   "Bivariate Exponential Distribution")))
> lambda_legend <- expression(paste(lambda[1],"=1/0.4"),
+                             paste(lambda[2],"=1/0.025"),
+                             paste(lambda[12],"=1/0.075"))
> legend(5,5,1,P33D$xyz.convert(18,0,12),legend = lambda_legend)
>
> lambda1 <- 50; lambda2 <- 50; lambda12 <- 10
> St1t2 <- exp(-t1/lambda1-t2/lambda2-pmax(t1,t2)/lambda12)
> P43D <- scatterplot3d(t1,t2,St1t2,highlight.3d=TRUE,
+   xlim=c(0,10),ylim=c(0,10),zlim=c(0,1),xlab=expression(t[1]),
+   ylab=expression(t[2]),zlab="S(t1,t2)",
+   main = expression(paste("A 3-d plot for ",
+   "Bivariate Exponential Distribution")))
> lambda_legend <- expression(paste(lambda[1],"=50"),
+                             paste(lambda[2],"=50"),
+                             paste(lambda[12],"=10"))
> legend(0,2,3,P43D$xyz.convert(18,0,12),legend = lambda_legend)
> dev.off()
null device
          1
```

Experimenting with different values of $\lambda_1, \lambda_2, \lambda_{12}$, the plots in Figure 2.2 are generated.

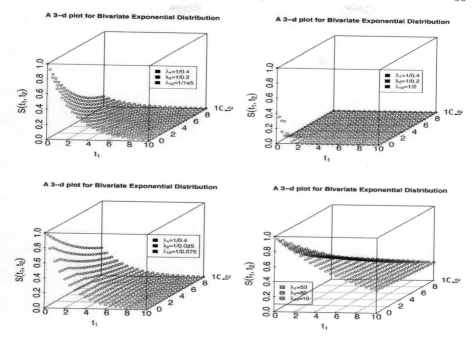

FIGURE 2.2: Bivariate Exponential

The joint pdf of (T_1, T_2) is:

$$f(t_1, t_2) = \begin{cases} \frac{1}{\lambda_1}\left(\frac{1}{\lambda_2} + \frac{1}{\lambda_{12}}\right)\exp\{-\frac{t_1}{\lambda_1} - \frac{t_2}{\lambda_2 + \lambda_{12}}\}, & \text{if } t_1 < t_2, \\ \frac{1}{\lambda_2}\left(\frac{1}{\lambda_1} + \frac{1}{\lambda_{12}}\right)\exp\{-\frac{t_2}{\lambda_2} - \frac{t_1}{\lambda_1 + \lambda_{12}}\}, & \text{if } t_1 > t_2, \\ \frac{1}{\lambda_{12}}\exp\{-\frac{t_2}{\lambda_1 + \lambda_2 + \lambda_{12}}\}, & t_1 = t_2. \end{cases} \qquad (2.5)$$

The marginal distributions obtained from $S(t_1, t_2)$ given in Equation 2.4 will be exponential distributions. The marginal distribution of T_1 is exponential with rate $\lambda_1 + \lambda_{12}$, and similarly for T_2, the parameter is $\lambda_2 + \lambda_{12}$ while the correlation between T_1 and T_2 is $(\lambda_1 + \lambda_2 + \lambda_{12})/\lambda_{12}$. However, the same can not be said about all forms of bivariate exponential distributions. For example, the PDF of Freund's bivariate exponential distribution is given by

$$f(t_1, t_2) = \begin{cases} \frac{1}{\lambda_1 \beta_2}\exp\left(-\frac{t_1}{\lambda_1 + \lambda_2 - \beta_2} - \frac{t_2}{\beta_2}\right), & \text{if } t_1 \leq t_2, \\ \frac{1}{\lambda_2 \beta_1}\exp\left(-\frac{t_2}{\lambda_1 + \lambda_2 - \beta_1} - \frac{t_1}{\beta_1}\right), & \text{if } t_1 \geq t_2. \end{cases} \qquad (2.6)$$

The marginal distributions of Freund's bivariate extension do not reduce to exponential. The survival function of this variant is given by

$$
S(t_1, t_2) \quad = \quad
\begin{cases}
1 - \frac{\lambda_1 + \lambda_2 - \beta_2}{\lambda_1} \exp\left(-\frac{t_1}{\lambda_1 + \lambda_2 - \beta_2} - \frac{t_2}{\beta_2}\right) \\
\quad - \frac{\lambda_1 + \lambda_2 - \beta_2}{\lambda_2 - \beta_2} \exp\left(-\frac{t_2}{\lambda_1 + \lambda_2}\right), \ \text{if } t_1 \leq t_2, \\
1 - \frac{\lambda_1 + \lambda_2 - \beta_1}{\lambda_2} \exp\left(-\frac{t_2}{\lambda_1 + \lambda_2 - \beta_1} - \frac{t_1}{\beta_1}\right) \\
\quad - \frac{\lambda_1 + \lambda_2 - \beta_1}{\lambda_1 - \beta_1} \exp\left(-\frac{t_1}{\lambda_1 + \lambda_2}\right), \ \text{if } t_1 \geq t_2.
\end{cases}
\tag{2.7}
$$

Balakrishnan and Lai (2009)[10] provides a treatise on bivariate lifetime distributions and chapter 10 of the book focuses on bivariate exponential distribution. Gumbel's Type I—III, various extensions of Freund's distribution, Hashino and Sugi, Block and Basu, Sarkar, etc. are discussed in the chapter with a study of the correlations as well as marginal distributions. Estimation techniques are also considered in the analysis. Nadarajah and Kotz (2006)[87] discuss an important problem in the context of bivariate exponential distribution and they highlight the important notion of reliability of bivariate system where $P(T_1 < T_2)$ is a critical probability. Nadarajah and Kotz derive the reliability for Gumbel's, Freund's, Hougaard's, Downtown's, Marshall and Olkin's, and Arnold and Strauss's bivariate exponential distributions.

Other distributions associated with bivariate lifetimes details, such as Weibull, can be found in Marshall and Olkin (2007)[74] and Balakrishnan and Lai[10] (2009). We consider other of extensions of lifetime distributions in the next section.

2.3   Generalized Lifetime Distributions

Weibull and gamma distributions are well known generalizations of the exponential distribution, as seen in any course on distribution theory or a first course in probability. Having a second parameter which determines the shape of the density function is one way of extending the exponential distribution. The simplest lifetime distribution can be extended in more ways too, and the two specific methods are through specification of *resilience* and *tilt* parameters. These two techniques can be applied on Weibull distribution too. While parametric distributions are not too popular in survival analysis, we would still recommend the reader to explore these possibilities before moving on to the semiparametric methods.

Marshall and Olkin (2007)[74], Chapter 9, record the first instance of the use of exponential distribution with resilience parameter in the period 1838–1845. It was found in the research of Verhulst and hence it is more appropriate to call it as *Verhulst distribution* instead of "exponential distribution with resilience parameter". The survival function of Verhulst distribution with a resilience parameter $\eta, \eta > 0$ is given by

$$S(t|\lambda,\eta) = 1 - \left(1 - e^{-t/\lambda}\right)^{\eta}, t, \lambda, \eta > 0. \tag{2.8}$$

The PDF and the hazard rate are respectively given in the following:

$$f(t|\lambda,\eta) = \frac{\eta}{\lambda}e^{-t/\lambda}\left(1 - e^{-t/\lambda}\right)^{\eta-1}, t, \lambda, \eta > 0, \tag{2.9}$$

$$h(t|\lambda,\eta) = \frac{\frac{\eta}{\lambda}e^{-t/\lambda}\left(1 - e^{-t/\lambda}\right)^{\eta-1}}{1 - \left(1 - e^{-t/\lambda}\right)^{\eta}}, t, \lambda, \eta > 0. \tag{2.10}$$

We will next generate the PDF, hazard rate, and survival functions for the Verhulst distribution with $\lambda = 1$ and the resilience parameters at $\eta = 0.25, 0.5, 0.75, 1.5, 5, 10$. Toward this, we first define the three functions in `Verhulst_PDF`, `Verhulst_SF` and `Verhulst_hr`, and then apply them over the time points 0.1, 0.2, ..., 10. The plot is obtained using the usual `plot`, `points`, `par`, and other R options. The result of the program is given in Figure 2.3.

```
> # Plotting the PDF, hazard rate, and
> # survival function of Verhulst Distribution
> Verhulst_PDF <- function(t,lambda,eta) {
+    eta*exp(-t/lambda)*(1-exp(-t/lambda))^(eta-1)/lambda
+ }
> Verhulst_SF <- function(t,lambda,eta){
+    1-(1-exp(-t/lambda))^eta
+ }
> Verhulst_hr <- function(t,lambda,eta){
+    (eta*exp(-t/lambda)*(1-exp(-t/lambda))^(eta-1)/lambda)/
+       (1-(1-exp(-t/lambda))^eta)
+ }
> t <- seq(0.1,10,0.1)
> eta_seq <- c(0.25,.5,.75,1.5,5,10)
> pdf("../Output/Verhulst_Distn_SF_PDF_HR.pdf",
+     height=10,width=25)
> par(mfrow=c(3,1))
> # The PDF
> plot(t,Verhulst_PDF(t,lambda = 1,eta=eta_seq[1]),"l",
+      ylim=c(0,1.5),xlim=c(0,7),xlab="Time",ylab="PDF",col=1)
> for(i in 2:length(eta_seq)){
+   points(t,Verhulst_PDF(t,lambda = 1,eta=eta_seq[i]),"l",col=i)
+ }
> legend(4,1.5,eta_seq,col=1:6,pch="_")
> title("The Verhulst Distribution")
> # The Survival Function
```

```
> plot(t,Verhulst_SF(t,lambda = 1,eta=eta_seq[1]),"l",
+      ylim=c(0,1),xlim=c(0,7),xlab="Time",ylab="S(t)",col=1)
> for(i in 2:length(eta_seq)){
+   points(t,Verhulst_SF(t,lambda = 1,eta=eta_seq[i]),"l",col=i)
+ }
> legend(4,1,eta_seq,col=1:6,pch="_")
> # The Hazard Rate
> plot(t,Verhulst_hr(t,lambda = 1,eta=eta_seq[1]),"l",
+      ylim=c(0,3),xlim=c(0,7),xlab="Time",ylab="h(t)",col=1)
> for(i in 2:length(eta_seq)){
+   points(t,Verhulst_hr(t,lambda = 1,eta=eta_seq[i]),"l",col=i)
+ }
> legend(4,3,eta_seq,col=1:6,pch="_")
> dev.off()
```

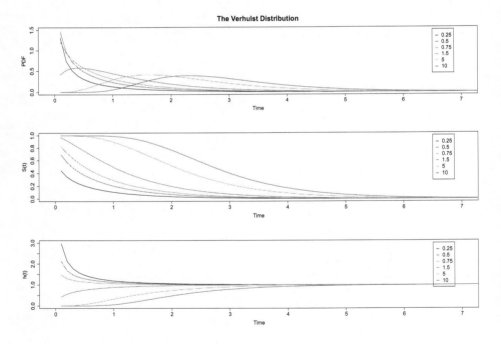

FIGURE 2.3: Verhulst Distributions

Fitting Verhulst distribution can be carried out using the MLE technique and the `fitdistr` function from the `MASS` package can be applied on the PDF `Verhulst_PDF`. Note that when we fit the exponential distribution to the Boeing AC data, it was observed that the simple distribution does not look suitable for all the aeroplanes. In the next example, we will fit the Verhulst distribution for the first aeroplane and then look at the AIC values for the exponential and Verhulst distribution for the tenth aeroplane data.

Example 3 *Fiting the Verhulst and Exponential Distributions for the Boeing AC Data. As stated earlier, we will use the* `fitdistr` *function for the first and tenth aeroplane data and obtain the output.*

```
> # Fitting Generalized Exponential Distribution
> t1_gexp <- fitdistr(t1,densfun = Verhulst_PDF,
+                      start=list(lambda=mean(t1),eta=1))
> t1_gexp
      lambda          eta
   95.6995145      0.9929147
  (26.6789653) ( 0.2772441)
> -2*t1_gexp$loglik + 2 # AIC
[1] 257.8128
>
> ## Fitting for Air Condition 10
> t10 <- as.numeric(na.omit(ACD[,10]))
> t10_exp <- fitdistr(t10,densfun = "exponential")
> t10_exp
       rate
   0.01219512
  (0.00304878)
> -2*t10_exp$loglik + 2 # AIC
[1] 175.015
> # Fitting Generalized Exponential Distribution
> t10_gexp <- fitdistr(t10,densfun = Verhulst_PDF,
+                      start=list(lambda=mean(t10),eta=0.5))
> t10_gexp
      lambda          eta
   56.3265823      1.8774206
  (16.1349724) ( 0.7036626)
> -2*t10_gexp$loglik + 2 # AIC
[1] 172.3814
```

For the first aeroplane, the fitted value of the resilience parameter is nearly equal to 1 which implies that the standard exponential distribution was suitable for the problem on hand. However, starting with an initial value of `eta=0.5` *for the resilience parameter, the fitted value is* **1.8774206** *and the AIC value is less compared with the AIC value for the fitted exponential distribution. Hence, we conclude that the effort of exploring the Verhulst distribution was suitable.*

□

The exponential distribution can be extended with a tilt parameter as well. Similarly, the Weibull distribution can be extended for the resilience parameters. The details of these type of extension can be obtained in Chapter 9 of Marshall and Olkin (2007)[74]. The bivariate exponential distribution has been extended to the resilience parameters in Kundu and Gupta (2009)[50], see `https://home.iitk.ac.in/~kundu/paper138.pdf`.

2.4   Censoring in Lifetime Studies

Early in the introductory chapter, the nature of survival data revealed the complications resulting from censoring and truncation. Unlike the Boeing data, we would rarely have access to datasets where every single observation would be complete. The sheep lifetimes, Subsection 1.2.1, were interval-censored, and we only know that the true lifetime for a sheep would be some value within the recorded interval. In case of data collected in Mayo Clinic's Primary Biliary Cirrhosis Study, Subsection 1.2.2, the time to death is the event of interest. Here, the event is observed for 161 patients while 232 were alive at the end of study and 25 required transplantation. The event has been censored for multiple reasons such as loss to followup, event not happening by the end of study, or a different event upending the observation. The incomplete observations are called as *censored survival times*. It is a common practice to call out the time to event as *survival time* even as that need not be the case. For instance, in certain other studies, the jailor might be observing the return of the prisoners for some crime after completely serving the last term, the banker might be following the customers for the time to default, while the beautician and barber might note the time to return of the favorite customers. As with the survival time, we will generally refer to the unit of observation in the subject as a patient. Thus, the four broad reasons for censoring are the following:

- the event of interest is not observed by end of the study

- the patient withdraws from the study

- the patient is lost to follow up

- a differnt event upends the follow up.

It is worthwhile mentioning that a patient in a clinical trial may experience severe adverse effects rendering the patient as "lost to follow up". When the observations are censored at the right-side, we call them out as *right-censored* and the mechanism is called as *right-censoring*. Aalen, et al. (2008)[2], Kalbfleish and Prentice (2002)[60], and Klein and Moeshberger (2003)[64] are a few of the benchmark books that can be referred for more details. How does censoring impact the analysis?

We will set up the likelihood function for an observed value. Let T continue to denote the failure time, or lifetime, of a patient and its distribution be F

with $F(t) = P(T \leq t) = \int_0^t f(s)ds$, where f is assumed to be absolutely continuous function. Of course, we continue to denote the survival probability with $S(t) = P(T > t) = 1 - F(t)$. Now, we will introduce the censoring time and denote it by C, and let $g(t)$ and $G(t)$ respectively be the PDF and CDF of the censoring RV. The observed data will be minimum of the lifetime (completed) and censored observation. Thus, if Y denotes the observed value, we have $\tilde{T} = \min\{T, C\} = T \wedge C$. We know that the observed value $\tilde{T}$ is a failure time if $T < C$ and we use the Kroneckar delta to represent this, $\delta = I(\tilde{T} = T) = I(T < C)$, explicitly,

$$\delta = \begin{cases} 1, & \tilde{T} = T, \text{ or } T < C, \\ 0, & \text{otherwise.} \end{cases}$$

We make an assumption that the lifetime and censoring distributions, F and G, are independent distributions. The likelihood contribution of a failure time, refer Section 1.4 of Duchateau and Janssen (2008)[37], is given by

$$
\begin{aligned}
L(\tilde{T} = T) = L(\tilde{t} = t, \delta = 1) &= \lim_{\epsilon \to 0} \frac{1}{2\epsilon} P\left(\tilde{t} - \epsilon < \tilde{T} < \tilde{t} + \epsilon, \delta = 1\right) \\
&= \lim_{\epsilon \to 0} \frac{1}{2\epsilon} P\left(\tilde{t} - \epsilon < \tilde{T} < \tilde{t} + \epsilon, \tilde{T} \leq C\right) \\
&\quad \text{the next step follows by applying} \\
&\quad \text{independence of } F \text{ and } G \\
&= \lim_{\epsilon \to 0} \frac{1}{2\epsilon} \int_{\tilde{t}-\epsilon}^{\tilde{t}+\epsilon} \int_t^\infty dG(c)dF(t) \\
&= \lim_{\epsilon \to 0} \frac{1}{2\epsilon} \int_{\tilde{t}-\epsilon}^{\tilde{t}+\epsilon} (1 - G(t))dF(t) \\
&= (1 - G(\tilde{T}))f(\tilde{T}).
\end{aligned}
$$

On similar lines, the likelihood contribution of a censored time is

$$L(\tilde{T} = C) = L(\tilde{t} = c, \delta = 0) = (1 - F(\tilde{t}))g(\tilde{t}),$$

and thus the likelihood function of an observed value $(\tilde{t}, \delta)$ is given by

$$L(\tilde{t}, \delta) = \left[(1 - G(\tilde{t}))f(\tilde{t})\right]^\delta \left[(1 - F(\tilde{t}))g(\tilde{t})\right]^{1-\delta}.$$

Given n independent observations $(\tilde{t}_i, \delta_i), i = 1, 2, \ldots, n$, the likelihood function is

$$L(\texttt{data}) = \prod_{i=1}^n \left[(1 - G(\tilde{t}_i))f(\tilde{t}_i)\right]^{\delta_i} \left[(1 - F(\tilde{t}_i))g(\tilde{t}_i)\right]^{1-\delta_i}. \qquad (2.11)$$

Apart from the assumption of independence of failure and censoring times, we need to make one more assumption. Specification of appropriate distribution for the censoring variable adds another problem besides censoring. Thus we make an additional assumption that the parameters of failure time distribution do not depend on the censoring mechanism and hence on G. This implies that the terms $(1 - G(\tilde{t}_i))^{\delta_i}$ and $g(\tilde{t}_i)^{1-\delta_i}$ do not influence inference related to F. This assumption is referred to as *noninformative censoring*. Dropping the terms, the likelihood function becomes

$$
L(\texttt{data}) = \prod_{i=1}^{n} (f(\tilde{t}_i))^{\delta_i} (S(\tilde{t}_i))^{1-\delta_i} = \prod_{i=1}^{n} \left(\frac{f(\tilde{t}_i)}{S(\tilde{t}_i)} \right)^{\delta_i} S(\tilde{t}_i) = \prod_{i=1}^{n} (h(\tilde{t}_i))^{\delta_i} S(\tilde{t}_i).
$$
(2.12)

Thus, we see how the hazard rate plays a natural role in analysis of censored data.

A few questions may arise naturally in the reader's mind. To keep the arguments simple, let us consider the case of noninformative censoring. In the case of right-censoring, instead of dealing with terms $S(\tilde{t}_i)$ for the cases with censoring, we can use a missing data technique—like the famous *Expectation-Maximization algorithm*, simply known as EM algorithm, and continue carrying out the inference using the classical inference techniques. In the next section, we will apply the EM algorithm for the interval censored data.

2.5   Handling Missing Data with EM Algorithm

The missing data problem is probably as old as data analysis itself. If the observations with missing values were not important—the analysis where all observations with missing values are deleted and analysis is carried out in the usual way is known as *complete case analysis*—we simply do not have the problem of dealing with missing data problem. As with the varied nature of censoring mechanisms, the theory of missing data mechanism is also richly developed. *Missing Completely at Random*, MCAR, is the case when the likelihood of a data point to be missing is completely random. Here, there is no reason for a data point missing with any values in the data set. It is missing because of pure chance reasons. *Missing at Random*, MAR, occurs when the likelihood of a missing data point is not related to the missing but it might be related to some of the observed values. In all other cases, the data are said to be *Missing Not at Random*, MNAR. Little and Rubin (1987, 2002)[73] gives a benchmark introduction to the three mechanisms, more so in the context of *survey data*. We will not delve deeper in these mechanisms and simply note that we will continue with the assumption that the missing value mechanism is MAR.

The general framework for handling missing data eluded for a long while and several attempts led to partial success. The EM algorithm debuted in the breakthrough research paper of Dempster, Laird, and Rubin (1977)[36]. A detailed account of the EM algorithm can be found in Little and Rubin (1987,

2002)[73] and McLachlan and Krishnan (1998, 2008)[80]. Schafer (1997)[99] details the development and application of EM algorithm for multivariate data. We will illustrate the EM algorithm on the sheep data set introduced in Section 1.2.1. An introduction of the general framework of the EM algorithm is in order before its application to our interest. We draw heavily from Section 16, Chapter 7, of Tattar, et al. (2016)[113].

We will denote the random vector by $\mathbf{X}$ and denote the realized value by $\mathbf{x}$. The realized values of $\mathbf{X}$ is referred to as the *observed data*. Let $\Omega_{\mathbf{X}}$ represent the sample space and the associated pdf is $g(\mathbf{x}, \theta)$, where $\theta = (\theta_1, \Delta\Delta\Delta, \theta_d)^T \in \Theta_{\mathbf{X}}$ is the vector of unknown parameters. We need to make a difference between missing data and incomplete data, and toward this we pretend that $\mathbf{X}$ is incomplete. The study consists of other variables, which are treated as *missing data* and we denote it by $\mathbf{U}$. Augmenting the missing data $\mathbf{U}$ with the incomplete data $\mathbf{X}$, we have the *complete data* denoted by $\mathbf{V}$ and we have $\mathbf{V} = (\mathbf{X}, \mathbf{U})$. The sample space of $\mathbf{V}$ is denoted by $\Omega_{\mathbf{V}}$ and the associated pdf by $g_c(\mathbf{v}, \theta)$. Thus, the likelihood of $\mathbf{v}$ is given by

$$\ln L_c(\theta|\mathbf{v}) = \ln g_c(\mathbf{v}, \theta).$$

Note that we have a many-to-one mapping of $\mathcal{V}$ to $\mathcal{X}$. In the standard EM notation we write $\mathbf{x} = \mathbf{x}(\mathbf{v})$ and hence we have the following relationship:

$$g(\mathbf{x}, \theta) = \int_{\Omega_V(x)} g_c(\mathbf{v}, \theta) d\mathbf{v}.$$

We begin with an initial value as the estimate of θ in $\theta^{(0)}$. Now, using the observed data $\mathbf{X}$ and the initial estimate $\theta^{(0)}$, we specify the conditional probability distribution $g_c(\mathbf{v}|\mathbf{x}, \theta^{(0)})$. Since the complete data loglikelihood $g_c(\mathbf{v}, \theta)$ is not observable, we replace it by its conditional expectation given $\mathbf{x}$ and $\theta^{(0)}$. This conditional expectation is the famous *Q-function*, the famous E-step of EM algorithm, and we now define it:

$$Q(\theta|\theta^{(0)}) = \int_{\mathcal{V}(\mathbf{x})} \ln L_c(\theta|\mathbf{v}) g_c(\mathbf{v}|\mathbf{x}, \theta^{(0)}) d\mathbf{v}.$$

The M-step maximizes $Q(\theta|\theta^{(0)})$ to obtain $\theta^{(1)}$ such that

$$Q(\theta^{(1)}|\theta^{(0)}) \geq Q(\theta|\theta^{(0)}).$$

The E-step and the M-step are repeated until convergence. We demonstrate the EM-algorithm for the sheep lifetimes data set in the next example.

Example 4 *Applying the EM Algorithm to the Sheep Lifetimes Dataset. Consider a random sample $T_1, T_2, \ldots, T_n$ of size n from an exponential distribution with failure rate λ as given in Equation 2.1, that is,*

$$f(t|\lambda) = \begin{cases} \frac{1}{\lambda} \exp\left(-\frac{t}{\lambda}\right), & t \geq 0, \lambda > 0, \\ 0, & otherwise. \end{cases}$$

In the case of interval censored data the exact values of the RVs T_i are not available and instead the available data is of the form $l_i \leq t_i \leq u_i, i = 1, \ldots, n$. Define $l = (l_1, \ldots, l_n), u = (u_1, \ldots, u_n)$, and $t = (t_1, \ldots, t_n)$. This form of data is common in clinical trials where the time to event of interest may occur between two consecutive scheduled visits. The observed data likelihood is then given by

$$L(\lambda | l, u) \propto \prod_{i=1}^{n} P(l_i \leq T_i \leq u_i) = \prod_{i=1}^{n} \{\exp(-l_i/\lambda) - \exp(-u_i/\lambda)\}.$$

In the missing data notation, we have $x_{obs} = (l, u)$ and $x_{mis} = t$. If t is actually observed, it is well-known that the maximum likelihood estimate, MLE, for the rate of exponential distribution is the sample mean $\hat{\lambda} = \bar{t}$. The log-likelihood function can be written as

$$\log(L_{mis}(\lambda | t)) = n \log\left(\frac{1}{\lambda}\right) - \frac{1}{\lambda} \sum_{i=1}^{n} t_i.$$

By the definition of conditional expectation, we get

$$E(T_i | l_i \leq T_i \leq u_i) = \lambda + \frac{l_i \exp(-l_i/\lambda) - u_i \exp(-u_i/\lambda)}{\exp(-l_i/\lambda) - \exp(-u_i/\lambda)}, i = 1, \ldots, n.$$

Now, let us denote the m-th iterate of parameter λ by $\hat{\lambda}^{(m)}$. Then, the E-step of the EM algorithm gives the expected value of the T_i's as

$$\hat{t}_i^{(m)} = \hat{\lambda}^{(m)} + \frac{l_i \exp(-l_i/\hat{\lambda}^{(m)}) - u_i \exp(-u_i/\hat{\lambda}^{(m)})}{\exp(-l_i/\hat{\lambda}^{(m)}) - \exp(-u_i/\hat{\lambda}^{(m)})}, i = 1, \ldots, n. \quad (2.13)$$

Using the $\hat{t}_i^{(m)}$s, the MLE of λ at the $(m+1)$-iteration as in the M-step is given by the average

$$\hat{\lambda}^{(m+1)} = \frac{\sum_{i=1}^{n} \hat{t}_i^{(m)}}{n}, m = 1, 2, \ldots. \quad (2.14)$$

The iterations are then continued until the convergence of $L(\lambda | l, u)$. In continuation of the R program, the EM algorithm for the interval censored data is given next. The function EM_Exponential_Interval_Censored *carries the necessary E- and M-steps. The E-step 2.13 is implemented in the function for the* xcurr *object and the M-step 2.14 in* lambda_imp.

```
> # Handling Missing Data with EM Algorithm
> data("Sheeplife")
> ux = lx = NULL
> for(i in 1:nrow(Sheeplife)){
```

```
+    lx[i] <- as.numeric(unlist(strsplit(as.character(
+        Sheeplife$Age_at_death[i]),"-")))[1]
+    ux[i] <- as.numeric(unlist(strsplit(as.character(
+        Sheeplife$Age_at_death[i]),"-")))[2]
+ }
> Lx <- rep(lx,Sheeplife$Frequency)
> Ux <- rep(ux,Sheeplife$Frequency)
> # Obtaining the likelihood function plot
> lik_fun <- function(lexp,uexp,lambda){
+    likfun <- sum(log((exp(-lexp/lambda)-exp(-uexp/lambda))))
+ }
> lambda_range <- seq(0,15,0.5)
> #plot(lambda_range,as.numeric(sapply(lambda_range,lik_fun,
> #      lexp=Lx,uexp=Ux)),"l",xlab=expression(lambda),
> #      ylab=expression(paste("L(",lambda,"|","x)")))
> lambda_range[which.max(as.numeric(sapply(lambda_range,
+        lik_fun,lexp=lx,uexp=ux)))]
[1] 6.5
> EM_Exponential_Interval_Censored <- function(xl,xu){
+    lexp <- xl; uexp <- xu
+    lambda_curr <- (mean(lexp+uexp))
+    lambda_imp <- lambda_curr+0.2
+    xcurr <- NULL
+    lambda_iter <- lambda_curr
+    loglik <- NULL
+    lambda_iter <- NULL
+    loglik <- NULL
+    iter <- 0; delta <- 0.5
+    while(abs(delta)>1e-4){
+      loglik_old <- sum(log(exp(-lexp/lambda_curr)
+                            -exp(-uexp/lambda_curr)))
+      xcurr <- lambda_curr + (lexp*exp(-lexp/lambda_curr)-
+                              uexp*exp(-uexp/lambda_curr))/
+        (exp(-lexp/lambda_curr)-exp(-uexp/lambda_curr))
+      lambda_imp <- (mean(xcurr))
+      loglik_new <- sum(log(exp(-lexp/lambda_imp)
+                            -exp(-uexp/lambda_imp)))
+      delta <- loglik_old-loglik_new
+      loglik <- c(loglik,sum(log(exp(-lexp/lambda_curr)
+                            -exp(-uexp/lambda_curr))))
+      iter <- iter + 1
+      lambda_iter <- c(lambda_iter,lambda_curr)
+      lambda_curr <- lambda_imp
+    }
+    return(cbind(1:iter,lambda_iter,loglik))
```

```
+ }
> EM_Exponential_Interval_Censored(Lx,Ux)
        lambda_iter     loglik
[1,] 1    14.526998 -1261.425
[2,] 2     7.236564 -1170.703
[3,] 3     7.209576 -1170.699
```

The function `EM_Exponential_Interval_Censored` *contains the computations related to* $\hat{t}_i^{(m)}$ *and* $\hat{\lambda}^{(m)}$) *as given in the earlier equations and the iterations are stopped when the change in the likelihood function* $L(\lambda|\boldsymbol{l}, \boldsymbol{u})$ *for two consecutive iterations become lesser than* `1e-4`, *or in plain numbers 0.0001.* □

Tempting it might be to use the EM algorithm to treat the censored observations as incomplete and carry out the inference. However, it is not advisable to substitute the censored observations by their expected values. There are two reasons for it. First, more often, we have to resort to nonparametric and semiparametric techniques in survival analysis. Thus, the Q-function is not easily computable. The second reason is that though the lifetimes might be censored, we have the information that the subject was alive till the censored time and EM algorithm can be used to handle missing data on covariates.

2.6   Counting Process Approach to Survival Analysis

Statistical inference related to the parametric distributions relies on strong assumptions on specifiction of the form. In the presence of censoring and truncation, the validation becomes tougher. As will be seen in the later chapters, Chapters 5–7, the semiparametric and nonparametric regression methods are more suitable, in general, for survival data. Even as the heuristic proofs might be acceptable for generic functions such as survival function and the cumulative hazard function, we need a framework which helps with asymptotics and inference. Aalen achieved the same in his doctoral thesis and this forms the content here.

We will continue to use the earlier setup of $T_1, T_2, \ldots, T_n$ denoting the time to event for these observations subject to, respectively, censoring variables $C_1, C_2, \ldots, C_n$. The observed variables are $\tilde{T} = \min\{T_i, C_i\}, \delta_i = I\{T_i \leq C_i\}, i = 1, 2, \ldots, n$. The sample $\tilde{T}_1, \tilde{T}_2, \ldots, \tilde{T}_n$, thus, contains some complete observations and some incomplete. Without loss of generality, we assume that there are no ties in the data and that the $\tilde{T}$s are ordered observations.

Now, let n_j and d_j respectively denote the number of individuals alive just before time $\tilde{T}_j$ and the number of events (deaths) at time $\tilde{T}_j$. The well-known

Nelson-Aalen estimator of the cumulative hazard function is

$$\hat{H}(t) = \sum_{j:\tilde{T}_j \leq t} \frac{d_j}{n_j},$$

and the famous *Kaplan-Meier estimator* of the survival function is

$$\hat{S}(t) = \prod_{j:\tilde{T}_j \leq t} \frac{n_j - d_j}{n_j}.$$

The derivation of the estimators will be taken up in Chapter 3. The incomplete observations and the product form of the estimator makes it challenging to establish the asymptotic properties. We need the *counting process framework* to derive the estimators and as a preliminary step, we will define it next.

The i-th counting process for observation i at time t is defined by

$$N_i(t) = I\{\tilde{T} \leq t, \delta_i = 1\}, i = 1, \ldots, n,$$

and we further define the aggregated counting process by

$$N(t) = N_\bullet(t) = \sum_{i=1}^{n} N_i(t).$$

Thus, at any time t, the counting process $N_i(t)$ tells us whether the event was observed for that observation, and $N = N_\bullet$ gives the overall number of events. We will drop the notation $N_\bullet$ in favor of N. The *at-risk process* associated with observation i and the overall at-risk process are given by

$$Y_i(t) = I\{\tilde{T}_i(t) > t\}, i = 1, \ldots, n,$$
$$Y(t) = Y_\bullet(t) = \sum_{i=1}^{n} Y_i(t).$$

Again, the $Y_\bullet$ will not be used any further. Thus, Y corresponds to the n_j's and $N(t) - N(t-)$ to the d_j's. In the formal *stochastic calculus* terms, observed times $\tilde{T}_i$'s are the stopping times of the counting process N. The general purpose of N is to count the total number of events occurring up to a given time instant, and measurability of the process holds with respect to the following filtration:

$$\mathcal{F} = \{\mathcal{F}_t = \sigma\{N_i(t), 0 \leq s \leq t, i = 1, 2, \ldots, n\}\}.$$

A brief note. It is unlikely for the reader with non-measure theory background to follow the need of a probability space, let alone a filtration. It is fine to skip over these details.

The *intensity process* associated with the counting process N is $Y(t)h(t)$, and the *cumulative intensity process* is $A(t) = \int_0^t Y(s)h(s)ds$. The cumulative intensity process has a different property from the counting process. While the counting process is a step function, jumping by 1 typically, it is also a stochastic process. We can not say whether $N(t)$ will change the value at time $t-$. On the other hand, the cumulative intensity process $A(t)$ value is known at time $t-$ and hence it is called as a *predictable process* too. Note the subtle difference between $A(t)$ and $H(t)$ where the latter is cumulative hazard function. When there are no individuals at risk, that is $Y(t) = 0$, the cumulative intensity process $A(t)$ will be 0 whereas $H(t)$ is a parametric function and is not 0. The cumulative intensity process $A(t)$ is also known as the *compensator* associated with the counting process $N(t)$.

Now, theoretically $N - A$ will be a *martingale*, and this property is preserved at the stopping times $\tilde{T}_i$'s as ensured by the *Doob's optional stopping theorem*.

The concepts of counting process, compensator, and the corresponding martingale is illustrated using the open-source software R. The following program is a simulation study where lifetimes are generated, the number of lifetimes computed, and the theoretical value of compensator at the lifetimes is deduced.

```
> # Understanding Martingale
> n <- 1e3
> Failure_rate <- 0.05
> set.seed(12345)
> Failure_obs <- rexp(n,rate=Failure_rate)
> Failure_obs <- sort(Failure_obs)
> Nt <- 1:n
> Yt <- n:1
> At <- c(0,-log(1-pexp(Failure_obs,Failure_rate)))
> dAt <- diff(At)
> Ct <- cumsum(Yt*dAt)
> X11(height=6,width=9)
> plot(Failure_obs,Nt,"l",xlab="Time",
+         ylab="Counting Process and the Compensator")
> points(Failure_obs,Ct,"l",col="red")
> legend(100,500,bty="n",lty=1,
+         c("Counting Process","Compensator"),
+         col=c("black","red"))
```

The program begins with $n = 1000$ observations. We generate n observations from exponential distribution with failure rate, say $\theta = 1/20 = 0.05$, by using the R function `rexp`. Here, `set.seed` is used to ensure the reproducibility. The failure times can be assumed to be in increasing order without loss

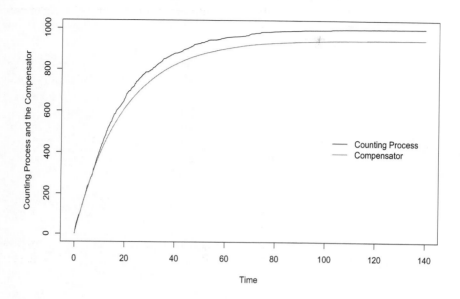

FIGURE 2.4: Understanding Martingale

of generality and the simulated observations are stored in `Failure_obs`. The counting process $N(t)$ is captured in `Nt` and $Y(t)$ in `Yt`. The expression $N(t)$ is very easy to obtain whereas $A(t)$ needs explanation. We need to evaluate the integral $A(t) = \int_0^t Y(u)\lambda(u)du$. The at-risk process Y has jump points at the observed times, and it is decreasing from n at time 0 to 0 after the last observation is realized. Thus, the steps of `At` (capturing the cumulative hazard function), `dAt`, and `Ct` perform the associated computation. The result of the R program is Figure 2.4. The program is applicable for complete observations, and it can be extended for the censored observations.

The purpose of using the counting processes is to put the asymptotics in the martingale theory framework which allows us to establish uniform consistency and asymptotic normality by application of Robelledo's martingale central limit theorem. This facilitates inference relating to the the compensator, which is the cumulative hazard function in the current discussion. Importantly, the estimators of Nelson-Aalen and Kaplan-Meier will be obtained first using the counting process framework and compensators, and then applying the limit theorems to their respective martingales, we will establish their useful properties. This task will be taken up in Chapter 3.

2.7   Multi-state Models

The time-to-event analysis studied up to this point is the waiting time for the occurrence of an event of interest. In some studies, the interest is in more than a single event as seen in a few studies in Section 1.2. For instance, in the study of Chronic Granulotomous Disease, Section 1.2.3, the infection is removed or treated, and observed for the recurrence of the infection and treated again. Thus, the number of events which can occur is more than one. The observation is also terminated if death occurs. Clearly, we have three states here namely infected state, infection-free state, and death. This is an example of *recurrent event*.

On the other hand, we observe in the study of Bone Marrow Transplant Data for Leukemia in Section 1.2.4, after the complex operation on bone marrow transplant is carried out, the transplantation is considered as failure if the leukemia relapses or death occurs within the remission time. A successful transplantation has two possibilities. The patient might see recovery of the platelet counts and then suffers from the graft-versus-host-disease (GVHD), or might see GVHD first and then followed by recovery to healthy platelet counts. The different states a patient can be observed are transplant, GVHD, platelet, relapse, and death. Here, the patient is stopped following the occurrence of either relapse or death, and hence we have two competing risks.

In Clinical Trials (CT) relating to chronic diseases such as cancer/AIDS, it is to determine a treatment which is most preferable among the available ones. It has been observed over the years that the natural endpoint considered in general, namely, the survival time is not conclusive enough for the problem on hand. Probability of Being in Response (PrBF), Quality Adjusted Life Years (QALYs), Time Without Symptoms of Disease and Toxicity (TWiST), and Quality Adjusted Lifetime (QAL) have emerged as some of the important endpoints. We will call CT involving QAL as a Health Related Quality of Life (HRQoL) study. In a HRQoL study, one classifies the health status of an individual into "health state" through the use of an appropriate questionnaire such as European Organisation for Research and Treatment of Cancer EORTC QLQ-30, or Functional Assessment of Cancer Therapy, FACT, or a generic questionnaire like Medical Outcome Study Short-Form Health Survey, MOS SF-36. To keep it simple, it suffices to say that a patient identified with a chronic disease is treated and then he goes through various health states before the eventual death. Of course, an observation may also get censored. Overall, we need to extend the time-to-event analysis and toward this we use the *multi-state model* structure. In a multi-state model, we have the state-space $\Gamma = \{$'State 1', 'State 2', ..., 'State k', $0\}$. For simplicity, we will denote the state space by $\Gamma = \{1, 2, \ldots, k, 0\}$. Note that though we used the same symbol Γ for the gamma integral, the context will make it clear that whether we are talking about the integral function or the state space. We will digress a bit and see how the digraphs look for the different class of multi-state models.

In the competing risks model, the present state is the alive state and death of a patient may occur on account of one of the causes. Thus, the transition happens from the alive state to death by one of the causes. Using the ggm package, we obtain the necessary digraph. The x- and y- coordinates are specified first. Using the DAG function, we specify the possible paths from the alive state to the cause of death. The rest of the program in the following code block is straightforward to follow.

```
> # Markov Models Digraphs
> ## Competitive Risks Model
> xCRM <- c(90,10,90,90,90)
> yCRM <- c(70,40,50,30,10)
> CRM_DAG <- DAG(Cause1~Alive,Cause2~Alive,
+                Cause~Alive,Causek~Alive)
> pdf("../Output/CRM_Digraph.pdf",height=50,width=50)
> drawGraph(CRM_DAG,coor=cbind(xCRM,yCRM))
> title("Digraph of Competing Risks Model")
> dev.off()
null device
          1
```

The result of using DAG and drawGraph function is the graphical display which is reproduced here in Figure 2.5.

The HRQoL study comes closest to the generic multi-state model. Here, death is the only absorbing state for the multi-state model and transitions are possible among all other possible states in any order. The method of plotting the CRM digraph can not be applied over here, and we need to define a matrix first. The numbers in the matrix are not really important. Using the markovchain package and the regular plot function, we are able to produce the digraph of the HRQoL state space. The result of the following code block is Figure 2.6.

```
> ## Health Related Quality of Life
> # States = [Pre,Sympton 1,Sympton 2,Symptom k,Death]
> TPM <- matrix(c(0.2,0.5,0.2,0,0.1,
+                 0.2,0.2,0.2,0.1,.3,
+                 0.2,0.2,0.2,0.1,.3,
+                 0.1,0.2,0.2,0.2,.3,
+                 0,0,0,0,1),nrow=5,byrow=TRUE)
> HRQoL <- new('markovchain',
+   transitionMatrix=TPM,
+   states=c('Pre','Symptom 1','Symptom 2','Symptom k','Death'))
> layout <- matrix(c(0,0,5,5,20,5,15,-5,25,0),ncol=2,byrow=TRUE)
```

Digraph of Competing Risks Model

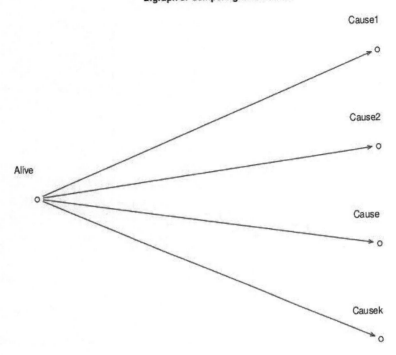

FIGURE 2.5: Competing Risks Model Digraph

```
> pdf("../Output/HRQoL_Digraph.pdf",height=50,width=50)
> plot(HRQoL,layout=layout,edge.arrow.size=1,
+       vertex.size=40,
+       main="HRQoL Markov Chain Digraph")
> dev.off()
null device
          1
```

Let $X(t)$ be a function that identifies the state of the patient at time $t, t \in \mathbb{R}^+$. We assume that $X(t)$ follows a nonhomogeneous Markov process with finite state space $\Gamma = \{1, 2, \ldots, k, 0\}$. The associated transition probability matrix, TPM, over the interval $(s, t]$, is $\mathbf{P}(s, t) = \{P_{jj'}(s, t)\}$, and the interpretation of $P_{jj'}(s, t)$ is that it gives the probability of starting at time s in state j and ending with state j' at time t. For the state space Γ, we define the *transition rate* from state j to state j' by

$$h_{jj'}(t) = \lim_{s \downarrow 0} \frac{P\left(X(t) = j' | X(t-s) = j\right)}{s}, \forall j, j' \in \Gamma. \tag{2.15}$$

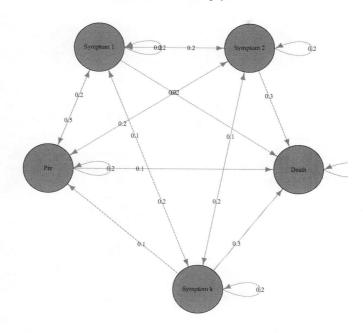

FIGURE 2.6: HRQoL Digraph

The *cumulative intensity matrix* is given by $\mathbf{H} = \{H_{jj'}\}$, where $H_{jj'}(t) = \int_0^t h_{jj'}(s)ds$. The TPM $\mathbf{P}(s,t)$ and $\mathbf{H}(t) = \{H_{jj'}(t)\}$ are related by

$$\mathbf{P}(s,t) = \prod_{(s,t]} \left(\mathbf{I} + d\mathbf{H}(u)\right), \tag{2.16}$$

Here, $\prod_{(s,t]}$ denotes the product integral of the integrand over the interval $(s,t]$, see ABGK.

The statistical inference related to $\mathbf{P}$ and $\mathbf{H}$ will be considered in Section 3.4.

2.8   Exercises

Exercise 2.1 *Prove the memoryless property of the exponential distribution, that is, for all $t, s \geq 0$, we have $P(T > t + s | T > s) = P(T > t)$.*   □

Exercise 2.2 *Series and Parallel System of Independent Units. In reliability analysis, it is common to have a system of units and the structure of the arrangement will determine the functioning of the system. The two common, and simpler, forms of the system are series and parallel structures. Assume that we have n identical units in the structure. A series system functions if all the units are functioning while a parallel system functions if even a single unit is working. Let the lifetime of the i-th component be denoted by $T_i, i = 1, \ldots, n$, and $S_i(t)$ denote the survival function of the component. Show that the survival of a series system at time t is*

$$S(t) = \prod_{i=1}^{n} S_i(t),$$

and for a parallel system

$$S(t) = 1 - \prod_{i=1}^{n} (1 - S_i(t)).$$

□

Exercise 2.3 *Write an R program to reproduce the survival plots of Weibull and gamma distribution as in Figures 1.2 and 1.4.*   □

Exercise 2.4 *Lifetime distributions can be fitted using the option of* densfun *in the* fitdistr *function. Explore fitting Weibull and gamma distributions to the first airplane airconditioning data. Which distribution is more apt based on the AIC?*   □

Exercise 2.5 *The benefit of using an object oriented programming such as R is the ability to create scalable code. Write an R program which will fit multiple lifetime distributions to the data related to all the ten aeroplanes.*   □

Exercise 2.6 *PDF and Survival Functions of Bivariate Exponential Distributions. Obtain the three-dimensional plots for the PDF and survival functions for the Freund's variant of bivariate exponential distribution respectively given in Equations 2.6 and 2.7.*   □

Exercise 2.7 *Stacy distribution is the generalization of gamma distribution as well as Weibull distribution. The pdf of the Stacy distribution is given by*

$$f(t|\lambda, \beta, \nu) = \begin{cases} \frac{\nu t^{\nu\beta-1} \exp(-(t/\lambda)^{\nu})}{\lambda^{\nu\beta}\Gamma(\nu)}, & t > 0, \nu, \lambda, \beta > 0, \\ 0, & otherwise. \end{cases}$$

Obtain the survival function of the Stacy distribution. □

Exercise 2.8 *Plot the Likelihood Function for a Single Observation. Suppose the lifetime variable denoted by T follows Weibull distribution with unknown parameters and the value of a realized observation is 13.9856. Plot the likelihood function for the cases of $\delta = 1$ and then repeat it for $\delta = 0$. State your inference.* □

Exercise 2.9 *Suppose that failure time T follows exponential distribution with failure rate λ_1 and the censoring time C follows exponential distribution with failure rate λ_2 with $\lambda_2 > \lambda_1$. Under the standard setup of right censoring and independence of failure and censoring times, what is the probability that the observed time will be a failure time?* □

Exercise 2.10 *In continuation of the previous example, suppose that the censoring time C is uniform over the interval $[a,b]$, $b > a > 0$. What is the probability of the observed time being failure time? For simplicity, take $\lambda_1 = 0.05$ and $a = 15, b = 40$.*

Hint: Running a simulation program in R might be easier over the analytical derivation. □

Exercise 2.11 *Simulated n interval censored lifetimes in two steps. First simulate the observations from an exponential distribution with rate λ, say 20. Put the observations in the interval bin 0–5, 5–10, ... Run the EM algorithm using the funciton* `EM_Exponential_Interval_Censored`. *Record your observations by varying the bin widths.* □

2.9 Más Lejos Temas

Statisticians Norman Johnson and Samuel Kotz have written a series of reference books on statistical distributions covering discrete, continuous, and multivariate distributions. The first series were released in the period 1969 to 1972. The successive editions comprehensively cover lifetime distributions and it remains the benchmark reference on distributions. An extension of the lifetime distribution of much importance is in the class of *frailty distributions*. In the context of discrete distributions, it happens that though the samples are drawn from same family of distributions, the observations are drawn from two or more distributions. Consequently, a single fitted probability distribution is not useful and one needs to deploy the notion of *mixed distributions*. In simple words, there is more variation in the data than what can be plausibly explained by a single probability distribution. In the context of lifetime distributions, the extension which best accomplishes the objective is *frailty distributions*. More details can be found in Duchateau and Janssen (2008)[37].

We had a peek into the type of censoring in Section 2.4 and as stated there censoring is assumed to be noninformative. In the case of informative censoring, the covariates influence the lifetimes and the censoring times too. Robins and Rotnitzky (1992)[96] provide an alternative way of estimating the survival functions and other parameters of interest through the concept of *inverse probability of censoring weights*, IPCW.

Exposure to counting process in Section 2.6 has been minimalistic. The reader will benefit with a detailed exposition as provided in Chapters 1–2 of Aalen, et al. (2008)[2]. Fleming and Harrington (1991)[44] and Andersen, et al. (1993)[7] also bring out lucidly the role of stochastic calculus that is adaptable in the context of survival analysis. Apart from Aalen, et al's 2008 work, an earlier emphasis on the point processes with an application to survival data can be found in Karr (1986)[62].

Chapter 3

Inference—Estimation

3.1   Introduction

The assumptions of parametric distribution are rigid and their validation becomes even tougher in the presence of censoring. Also, though the principal of MLE invariance preserves the asymptotic properties, the transformation from properties of the parameters to functions looks not quite straightforward, especially for multi-parameter distributions. The quantities of interest in clinical trials is often the survival function and the cumulative hazard function. Though the functional delta theorem is useful in establishing the asymptotic properties, the nonparametric methods of estimating these quantities is appealing. Thus, we undertake the nonparametric method of estimating the survival function and cumulative hazard function. The powerful techniques of Kaplan-Meier estimator and Nelson-Aalen estimator are introduced in Section 3.2, and their extension to carry out inference related to other statistical functions of interest, such as the mean and median survival times is dealt with.

Besides survival function and cumulative hazard function the hazard rate is also of interest and sometimes the practitioner might need it. The problem of estimating the hazard rate is similar to that of estimating the probability density function. We need smoothing techniques to transform the cumulative hazard function into estimates of the hazard rate. In Section 3.3, we use the kernels to accomplish this task.

The multi state situation lends itself, naturally, to modeling through a Markov process. It would however be necessary to assume that, given the present state, transition from this state to another in one observation instance is not influenced by the states occupied by the patient prior to the present one, which amounts to assuming the Markov process. Further, given the nature of a study like a clinical trial, assumption of time homogeneity is untenable. We thus proceed to carry out inference under a nonhomogeneous Markov process model. The transition probability matrix (TPM) of the assumed nonhomogeneous Markov process plays a pivotal role in understanding multi-state models. The estimation of the TPM is undertaken in Section 3.4. The sojourn time distribution in each transitive state is also taken up here.

DOI: 10.1201/9781003306979-3

3.2   Nonparametric Estimation

We begin the section with estimation of the cumulative hazard function $H(t) = \int_0^t h(s)ds$. The estimator of the cumulative hazard function $H(t)$ for right-censored data was first obtained by Nelson in 1969. Nelson used the then standard nonparametric methods and derived the estimator and gave a heuristic proof of the asymptotic properties of the estimator. Aalen independently derived the same estimator in the stochastic calculus framework and firmly established the asymptotic properties of uniform consistency and asymptotic normality. Aalen established the results in his 1975 PhD thesis, Aalen (1975)[4]. The nonparametric estimator is since referred to as *Nelson-Aalen estimator* and we simply denote it by $\hat{H}(t)$.

It might be tempting to use $\hat{H}(t)$ as the base for the estimation of the survival function $S(t)$ with an application of the functional delta theorem. However, estimation of the survival function using heuristic estimator is rich in intuition and arises quite naturally. Kaplan and Meier (1958)[61] proposed the estimation of the survival function for right-censored data using heuristics, and the resulting estimator takes the form of a product of terms and thereby earning it the moniker of *product-limit estimator*. The survival function estimator $\hat{S}(t)$ obtained in the 1958 paper is popularly known as the *Kaplan-Meier estimator*. The proof outlined by the inventors was again heuristic and the definitive (asymptotic) properties were once again proved in Aalen's PhD thesis. The framework continues to be the stochastic calculus framework. The Kaplan-Meier estimator is one of the path-breaking works in the broader subject of Statistics and it is considered as one of the twenty important contributions of the twentieth century. The variance of the Kaplan-Meier (KM) estimator is given by the Greenwood formula and it predates the KM estimator. We will now begin by restating the usual notations.

We will continue to denote the time to event of interest, from Section 2.4, for the n individuals by $T_1, T_2, \ldots, T_n$, and the censoring variables with $C_1, C_2, \ldots, C_n$. The observed data is the n pair of observations $(\tilde{T}_i, \delta_i), i = 1, 2, \ldots, n$, with $\tilde{T}_i = T_i \wedge C_i$ and $\delta_i = I(\tilde{T}_i = T_i)$. Our goal is estimation of the cumulative hazard function, $H(t)$. Without loss of generality, going forward and rest of the book we will assume $\tilde{T}_1 \leq \tilde{T}_2 \leq \ldots \leq \tilde{T}_n \leq \ldots$. Let $N(t) = \sum_{i=1}^n N_i(t)$ be the associated counting process, where $N_i(t)$ captures if the event is observed for the i-th data point by time t, and further let $Y(t) = \sum_{i=1}^n Y_i(t)$ be the at-risk process. We will now derive an estimator for the cumulative hazard function $H(t)$. The developments in this chapter rely heavily on Aalen, et al. (2008)[2].

3.2.1 Nelson-Aalen Estimator

We will begin with a *single observation*. Suppose that it is known that at time t the event has not yet occurred and we would like to find probability of the event, say 'alive' to 'dead', occurring in the interval $[t, t + dt)$. By the definition of the hazard rate, we have

$$P(dN(t) = 1|\mathcal{F}_t) = h(t)dt,$$

where $\mathcal{F}_t$ is the σ-field generated by $N(t)$. Since $dN(t)$ is a binary variable and the hazard rate is a predictable process, we can rewrite the previous equation in terms of expectations as follows:

$$E(dN(t) - h(t)dt|\mathcal{F}_t) = 0.$$

Evaluating the expression, we get $N(t) - H(t) = 0$, and for the process $M(t)$ defined by $M(t) = N(t) - H(t)$, we can also see that

$$E(dM(t)|\mathcal{F}(t)) = 0.$$

The increment in the counting process written as $dN(t) = h(t)dt + dM(t)$, the *stochastic differential equation* of interest, has the classical interpretation of 'observation = signal + noise', see page 27 of Aalen, et al. (2008), with the observation represented by $dN(t)$, $h(t)dt$ the signal, and $dM(t)$ the noise.

When the observation T is subject to a random censoring time C, the observed data is $\tilde{T} = T \wedge C, \delta = I\{T \leq C\}$. The probability of the counting process $N(t)$ observing the event in the interval $[t, t + dt)$ is then given by

$$P(dN(t) = 1|\mathcal{F}_t) = \begin{cases} h(t)dt, & T \geq t, \\ 0, & \texttt{otherwise}, \end{cases}$$

and we can thereby rewrite the compensator $dH(t)$ in terms of the hazard rate and the "at-risk" process as follows:

$$dH(t) = h(t)I(T \geq t)dt.$$

This is the well-known *multiplicative intensity model*.

With n observations, we can summarize the history of events through the filtration:

$$\mathcal{F}_t = \sigma\{N_i(t), 0 \leq s \leq t, i = 1, 2, \ldots, n\}.$$

While $\mathcal{F}_t$ summarizes the history of occurrences of the event up to and at time t, the past of the event history just prior to time t is captured by $\mathcal{F}_{t-}$ and it is defined by

$$\mathcal{F}_{t-} = \sigma\{N_i(t), 0 \leq s < t, i = 1, 2, \ldots, n\}.$$

Thus, the filtration $\mathcal{F}_{t-}$ captures the "past". The aggregated process is then

$$N(t) = \sum_{i=1}^{n} N_i(t) = \sum_{i=1}^{n} I\{\tilde{T} \leq t, \delta_i = 1\},$$

and the multiplicative intensity model is

$$dH(t) = h(t)Y(t)dt,$$

where $Y(t) = \sum_{i=1}^{n} Y_i(t) = \sum_{i=1}^{n} I\{T_i \geq t\}$. The stochastic differential equation with the aggregated counting process and the intensity model is thus given by

$$dN(t) = h(t)Y(t)dt + dM(t). \tag{3.1}$$

It is tempting to divide both sides of the above equation by $Y(t)$. However, there it is possible that at some time t, the at-risk process might become zero. We thus introduce a new indicator $J(t) = I(Y(t) > 0)$ and adopt the convention that $J(t)/Y(t) = 0$ whenever $Y(t) = 0$. Multiplying the Equation 3.1 by $J(t)$ and then dividing it by $Y(t)$, we get

$$\frac{J(t)}{Y(t)}dN(t) = J(t)h(t)dt + \frac{J(t)}{Y(t)}dM(t),$$

and integrating this expression over the interval $[0, t]$, we obtain

$$\int_0^t \frac{J(s)}{Y(s)}dN(s) = \int_0^t J(s)h(s)ds + \int_0^t \frac{J(s)}{Y(s)}dM(s). \tag{3.2}$$

We will evaluate the integral $\int_0^t \frac{J(s)}{Y(s)}dN(s)$ after considering the right-hand quantities first. Define $H^*(t) = \int_0^t J(s)h(s)ds$. Let us consider the second term in $\int_0^t \frac{J(s)}{Y(s)}dM(s)$. Because $Y(t)$ is a predictable process with respect to the filtration $\mathcal{F}_t$ and $J(t) = I(Y(t) > 0)$, we can easily see that $J(t)$ is also a predictable process and thereby the ratio $J(t)/Y(t)$ too. Thus, $\int_0^t \frac{J(s)}{Y(s)}dM(s)$ is a martingale and it's mean is also zero.

The integral $\int_0^t \frac{J(s)}{Y(s)}dN(s)$ is of interest to us. It follows from the standard results of *stochastic calculus* that when the variable of integral, $N(t)$ here, is a counting process, the stochastic integral is nothing but Lebesgue-Stieltje integral and the solution is simply the sum of the integrand, $J(s)/Y(s)$, over the jump points. Thus,

$$\int_0^t \frac{J(s)}{Y(s)}dN(s) = \sum_{T_j \leq t} \frac{J(T_j)}{Y(T_j)}.$$

Note that when we are evaluating $\frac{J(T_j)}{Y(T_j)}$ at the observed times, failure or censored observation, it is true that $Y(T_j) > 0$ and hence we can simplify the previous expression as follows:

$$\int_0^t \frac{J(s)}{Y(s)}dN(s) = \sum_{T_j \leq t} \frac{1}{Y(T_j)}.$$

The quantity $\sum_{T_j \leq t} \frac{1}{Y(T_j)}$ is just an estimate of the cumulative hazard function and it is the well-known *Nelson-Aalen estimator*:

$$\hat{H}(t) = \sum_{T_j \leq t} \frac{1}{Y(T_j)}. \tag{3.3}$$

Substituting the expressions for $\int_0^t \frac{J(s)}{Y(s)} dN(s)$ and $H^*(t) = \int_0^t J(s)h(s)ds$ in Equation 3.2, we get the following after a little rearrangement:

$$\hat{H}(t) - H^*(t) = \int_0^t \frac{J(s)}{Y(s)} dM(s). \tag{3.4}$$

The right-hand quantity of the previous equation is a zero-mean martingale, and hence $\hat{H}(t)$ is an unbiased estimator of $H^*(t)$. It is not possible to $H(t)$ when $Y(t) = 0$. The optional variation process of the martingale $\hat{H}(t) - H^*(t)$ is $\left[\hat{H} - H^*\right](t) = \int_0^t \frac{J(s)}{Y(s)^2} dN(s)$, and hence the variance of the Nelson-Aalen estimator 3.3 is calculated by

$$\hat{\sigma}^2(t) = \text{Var}(\hat{H}(t)) = \sum_{T_j \leq t} \frac{1}{Y(T_j)^2}. \tag{3.5}$$

The proof of the asymptotic distribution of the Nelson-Aalen estimator $\hat{H}(t)$ follows by application of the martingale central limit theorem on the martingale $\hat{H}(t) - H^*(t)$ and the fact that $H^*(t)$ and $H(t)$ are asymptotically equivalent. The standard assumption is that the process $Y(t)/n$ stabilizes for large n, equivalently $Y(t)/n \xrightarrow{P} y(t), \forall t \in [0, \tau]$, where τ is the largest time at which $h(t)$ is nonzero. For comprehensive details, the reader can refer Chapters 2 and 3 of ABG[2].

Using the standard arguments, a $100(1 - \alpha)\%$ confidence interval for $H(t)$ is given by

$$\hat{H}(t) \pm z_{1-\alpha/2}\hat{\sigma}(t).$$

We will next apply these techniques to the PBC dataset.

Example 5 *Estimating the Cumulative Hazard Function for the PBC Data. We continue the usage of the PBC dataset which had been introduced earlier in Section 1.2.2. We had remarked previously that the purpose of the clinical trial was to find if the drug D-penicillamine improves the lifetime of the patients treated for the PBC problem. Here, we have n = 418 patients in the study out of which 106 patients did not participate in the study but gave only basic measurements. In the randomized clinical trial, 158 patients received the drug D-penicillamine while the rest were administered the placebo. We first subset the dataset for the D-penicillamine treatment only, and obtain the cumulative hazard function $\hat{H}(t)$. Later, we will obtain the Nelson-Aalen estimator for the placebo, and finally consider overall patients in the study. Lets begin.*

```
> data(pbc)
> nrow((pbc))
[1] 418
> table(pbc$trt)
  1   2
158 154
> table(pbc$trt,exclude=NULL)
    1    2 <NA>
  158  154  106
> pbc.D <- subset(pbc,trt==1,select = c(time,status))
> dim(pbc.D)
[1] 158    2
> DSurv <- Surv(pbc.D$time,pbc.D$status==2)
> DSurvFit <- survfit(DSurv~1)
```

The data *function loads the dataset* pbc *from the* survival *package. We check on the number of observations using* nrow *and then verify the number of patients given the treatment and the placebo. Because there were missing observations in the dataset for the treatment, patients who agree to give only physical measurements, the* table *function does not add up to 418, and using the option of* exclude=NULL *we ensure completion of the information. The survival times and the censoring indicator for the treatment are subsetted in the data frame* pbc.D. *Though we use the* Surv *and* survfit *functions, we will use them to setup a different explanation.*

Using an object oriented software such as R, one readily computes the Nelson-Aalen estimator and the Kaplan-Meier estimator, and also comfortably plots the cumulative hazard function and the survival function. However, we need to understand the underlying formulas, and we will next see how the counting processes and the at-risk process are stacked in the R infrastructure.

The DSurvFit *object consists of the times and the associated failure time indicator and we will first inspect the data. The reader might wonder what was the necessity to fit the survival object on it and then display when the same could have been obtained in* pbc.D *data frame. The reason is that following the application of the survival function, the observed times $\tilde{T}s$ are arranged in chronological order. With the times arranged in the required order, it is straight-forward to obtain the number of events occuring at the times and also the at-risk processes. The cumulative sum of the events* cumsum(DSurvFit$n.event) *gives the counting process $N(t)$ while* DSurvFit$n.risk *readily gives the data for the at-risk process.*

```
> # Understanding the counting process framework
> pdf("../Output/Counting_AtRisk_Processes.pdf")
> par(mfrow=c(2,1))
> DSurvFit$time # the times at which events or censoring occur
  [1]   41   71  131  140     732  737
 [18]  750  762  799  824    1077 1083
```

```
[154] 4500 4556
> DSurvFit$n.event # indicator of the event occurring at time t
  [1] 1 1 1 1 1     0 0 1 0 0 1
 [45] 0 0 1 1 0     0 1 1 1 0 0

[133] 0 1 0             0 0 0 0 0
> cumsum(DSurvFit$n.event) # the counting process N(t)
  [1] 1  2  3  4  5        21 22
 [30] 23 24                36 37

[146] 63 63 64 65 65 65 65 65 65 65
> DSurvFit$n.risk # the at-risk process Y(t)
  [1] 158 157 156     139 138 137
 [23] 136 135 134     116 115 114

[133] 23 22 21     4  3  2  1
> # Plotting the counting process against the time
> plot(DSurvFit$time,cumsum(DSurvFit$n.event),"l",
+    xlab="Time",ylab="N(t)")
> # Plotting the at-risk process against the time
> plot(DSurvFit$time,DSurvFit$n.risk,"l",
+    xlab="Time",ylab="Y(t)")
> dev.off()
null device
          1
```

The result of the above code block is Figure 3.1. Obviously, the counting process $N(t)$ is an increasing process, and the at-risk is decreasing. However, the figure lays out the time points around which the events are occurring clearly. Certainly, using cumsum(DSurvFit$n.event) *and* DSurvFit$n.risk *one can write codes and obtain the Nelson-Aalen estimator as given in expression 3.3. However, we will use the fitted survival object and then using the values therein, we obtain the Nelson-Aalen estimator. The reason of this workaround is that unlike the Kaplan-Meier estimator, we do not outrightly obtain the Nelson-Aalen estimator in the software.*

The survival object DSurvFit *is an object of class* survfit *in R. Using the* summary *function, we extract more details of the fitted survival object. Next, using* -log(DSurvSumm$surv)*, we obtain the cumulative hazard function estimates at the jump times, which happens to be the times t at which $\delta = 1$. The estimates of the cumulative hazard function are stored in the object* DHaz*. To keep the tail smooth, we augment the time with a large time point, 6000 and also add the last observed cumulative hazard function value corresponding to the value of time at 6000 days.*

Survival Analysis

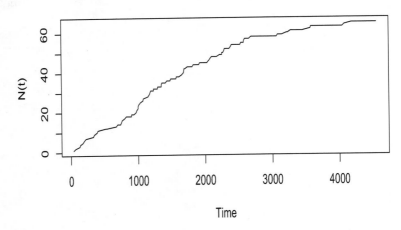

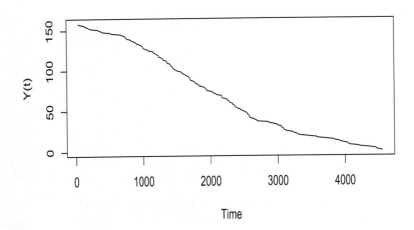

FIGURE 3.1: The Counting Process and the At-Risk Process

```
> DSurvSumm <- summary(DSurvFit)
> DHaz <- -log(DSurvSumm$surv)
> DTime <- c(DSurvSumm$time,6000)
> DHaz <- c(DHaz,tail(DHaz,1))
> plot(DTime,DHaz,xlab="Time",ylab="Cumulative Hazard",
+       main="Cumulative Hazard Plot","l")
```

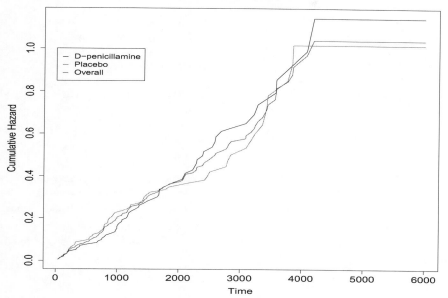

FIGURE 3.2: Nelson-Aalen Estimator for the PBC Data

The result of the above code block is black curve in Figure 3.2. We further subset the original data set for placebo, pbc.P, and repeat the computation related to the Nelson-Aalen estimator, and then extend it for the overall number of observations in the dataset. The resulting cumulative hazard function plots are the red and blue colored curves appended in the same figure.

```
> pbc.P <- subset(pbc,trt==2,select = c(time,status))
> PSurvSumm <- summary(survfit(Surv(pbc.P$time,
+     pbc.P$status==2)~1))
> PHaz <- -log(PSurvSumm$surv)
> PTime <- c(PSurvSumm$time,6000)
> PHaz <- c(PHaz,tail(PHaz,1))
> points(PTime,PHaz,"l",col="red")
> pbc.A <- subset(pbc,select = c(time,status))
> ASurvSumm <- summary(survfit(Surv(pbc.A$time,
+     pbc.A$status==2)~1))
> AHaz <- -log(ASurvSumm$surv)
> ATime <- c(ASurvSumm$time,6000)
> AHaz <- c(AHaz,tail(AHaz,1))
> points(ATime,AHaz,"l",col="blue")
> legend(0,1,c("D-penicillamine","Placebo","Overall"),
+         col=c("black","red","blue"),pch="_")
```

Upto time, approximately, 1700 days, the cumulative hazard curve for the D-penicillamine is below the placebo. The cumulative hazard curve exceeds from this time until around 3500 days, and then there are two more crossings of the cumulative hazard curves. We can not conclude overall which treatment to select based on the plots of Figure 3.2. We will inspect in the next chapter whether the two curves are significantly similar or dissimilar. □

Confidence intervals play an important part in completing the statistical inference. At a given time t, the $100(1 - \alpha)\%$ confidence interval is given by $\hat{H}(t) \pm z_{1-\alpha/2}\hat{\sigma}(t)$, and $\hat{\sigma}(t) = \left(\sum_{T_j \leq t} \frac{1}{Y(T_j)^2}\right)^{1/2}$. We will codify these formulas in the next example.

Example 6 *Confidence Intervals for the Cumulative Hazard Functions for the PBC Data.* *We need to calculate the variance first to setup the confidence intervals. Now, the expression* $\sum_{T_j \leq t} \frac{1}{Y(T_j)^2}$ *is aptly computed by using* DSurvSumm$n.event/DSurvSumm$n.risk^2 *and then accumulating them using the* cumsum *function. The rest of the program is easier to understand.*

```
> #variance decrements
> DDecVar <- DSurvSumm$n.event/DSurvSumm$n.risk^2
> DVar <- cumsum(DDecVar) # the variance of NA at time t
> DStdDev <- sqrt(DVar) # Standard Deviation
> DStdDev <- c(DStdDev,tail(DStdDev,1))
> alpha <- 0.05
> DUL <- DHaz + qnorm(1-alpha/2)*DStdDev # Upper limit
> DLL <- DHaz + qnorm(alpha/2)*DStdDev # Lower limit
> plot(DTime,DUL,xlab="Time",ylab="Cumulative Hazard",
+      main="Cumulative Hazard Plot","l")
> points(DTime,DHaz,"l",lwd=3)
> points(DTime,DLL,"l")
```

The upper and lower confidence limits are plotted with the Nelson-Aalen estimator. The result is Figure 3.3. The confidence interval is thus obtained for the D-penicillamine treatment. □

In many instances, it is complex to carryout inference, especially when the asymptotics are tougher to establish. The issue is not restricted to nonparametric inference, or semiparametric models as they will be discussed later, but for different functions of the unknown parametrics. Resampling and bootstrap technique provide a generic algorithm that can be used to carry out the inference without knowing the exact or asymptotic distributions. Efron (1979)[39]

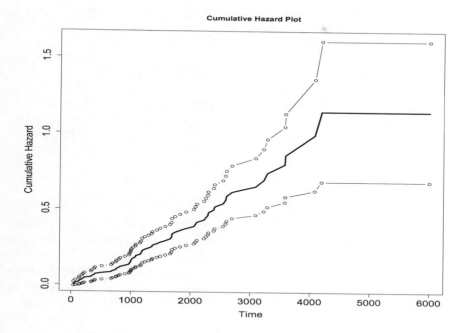

FIGURE 3.3: Confidence Intervals for the Nelson-Aalen Estimator

invented the bootstrap technique and we discuss the generic framework first and then extend it to the survival data.

Suppose $W_1, W_2, \ldots, W_n$ is a random sample from a distribution F, and the parameter of interest is θ. Let $T = g(W_1, \ldots, W_n)$ be an estimator of θ. In the bootstrap method, the process is repeated for a large number of times, say B. The bootstrap algorithm then unfolds as follows:

- Draw a sample of size n **with replacement** from $W_1, W_2, \ldots, W_n$ and denote it with $W_1^{*1}, W_2^{*1}, \ldots, W_n^{*1}$. This sample is called as *a bootstrap sample*.

- Obtain the estimate $T^{*1} = g(W_1^{*1}, \ldots, W_n^{*1})$ using the bootstrap sample.

- Repeat the previous step $B - 1$ number of more times to obtain $T^{*2}, T^{*3}, \ldots, T^{*B}$.

- Using the sampling distribution of $T^{*1}, T^{*2}, \ldots, T^{*B}$ to carryout the inference about θ.

A few remarks are in order here. The random variables W's need not be scalar and can be vector-valued too. The same can be said regarding θ. The sampling distribution can be used irrespective of the true probability distribution F, and we do not need to know the probability distribution of T at all.

As such there are no guidelines or rules over the number of replications B, and in most practical applications it has been found that $B = 50$ suffices. We note that the number of bootstrap samples is not related with the number of observations n.

When we draw n observations with replacement from n, there are bound to be repetitions. How many unique observations, on average, do we get? The answer is $0.632 \times n$, and hence the bootstrap algorithm is sometimes called as the 0.632 algorithm.

At the outset, it would be tempting to believe that the popular bootstrap technique can be straightaway applied to survival data, and we can readily use the bootstrap samples to obtain the sampling distribution of, say, $H(t)$ for some time t. We need to make some modifications to carry out the bootstrap technique on the survival data. We note that sampling with replacement is also at the index level. The Nelson-Aalen estimator $\hat{H}(t)$ has jump points at the failure times, and it remains flat between two consecutive failure times. Thus, the jump points of Nelson-Aalen estimator $\hat{H}(t)$ based on observed data $(\tilde{T}_1, \delta_1), (\tilde{T}_2, \delta_2), \ldots, (\tilde{T}_n, \delta_n)$ are going to be different from the bootstrap sample $(\tilde{T}_1^{*1}, \delta_1^{*1}), (\tilde{T}_2^{*1}, \delta_2^{*1}), \ldots, (\tilde{T}_n^{*1}, \delta_n^{*1})$, as well as the later ones with $(\tilde{T}_1^{*2}, \delta_1^{*2}), (\tilde{T}_2^{*2}, \delta_2^{*2}), \ldots, (\tilde{T}_n^{*2}, \delta_n^{*2})$, and so forth. In the next example, we will show how the bootsrapping is modified for the survival data.

Example 7 *Bootstrap Confidence Intervals for the Cumulative Hazard Functions for the PBC Data. We continue to explore the PBC dataset and further focus on the data related to the treatment data of D-penicillamine. In Example 5, we had obtained the Nelson-Aalen estimator for the treatment, and the confidence interval were also obtained in Example 6. Now, we will obtain the bootstrap confidence interval.*

The number of bootstrap iterations is fixed at B <- 50. *Now, we create a data frame with the time points at which the events occur for the patients under the treatment. In the loop, we store the bootstrapped data in the placeholder* pbc.B. *The simple* sample *function helps in carrying out the sampling with replacement. While the Nelson-Aalen estimator is easily obtained as earlier, we need to obtain the estimate of the* $H(t)$ *at the failure time points of original data. This is obtained by using the options of* times *and* extend=TRUE *options with the* summary *function.*

```
> # Bootstrapping the Nelson-Aalen Estimator
> B <- 50 # The number of  bootstrap samples
> # the times at which we need the Nelson-Aalen estimate of
> # the bootstrap sample
> c(head(DTime),tail(DTime))
 [1]   41   71  131  140  179  198 3282 3574 3584 4079
[11] 4191 6000
> BootDF <- data.frame(matrix(NA,nrow=length(DTime),ncol=B+3))
```

```
> colnames(BootDF) <- c("Time",paste0("BS_NA",1:B),"BUL","BLL")
> BootDF$Time <- DTime
> n <- nrow(pbc.D)
> for(i in 1:B){
+    set.seed(123+i)
+    pbc.B <- pbc.D[sample(1:n,n,replace=TRUE),]
+    BSurv <- Surv(pbc.B$time,pbc.B$status==2)
+    BSurvFit <- survfit(BSurv~1)
+    DSurvSumm <- summary(BSurvFit,times=DTime,extend=TRUE)
+    BootDF[,i+1] <- -log(DSurvSumm$surv)
+ }
> for(j in 1:nrow(BootDF)){
+    BootDF$BLL[j] <- quantile(BootDF[j,2:B+1],alpha/2)
+    BootDF$BUL[j] <- quantile(BootDF[j,2:B+1],1-alpha/2)
+ }
> pdf("../Output/PBC_NA_Bootstrap_Confidence_Intervals.pdf",
+     height=10,width=10)
> plot(DTime,DUL,xlab="Time",ylab="Cumulative Hazard",
+       main="Bootstrap Confidence Interval","l")
> points(DTime,DHaz,"l",lwd=3)
> points(DTime,DLL,"l")
> points(DTime,BootDF$BLL,"l",col="blue")
> points(DTime,BootDF$BUL,"l",col="blue")
> dev.off()
null device
          1
```

The rest of the program is straightforward. We plot the asymptotic confidence interval along with the bootstrap confidence interval. The result is Figure 3.4.

□

We will now move on to the problem of estimating the survival function $S(t)$.

3.2.2   Kaplan-Meier Estimator

Kaplan and Meier (1958)[61] makes a detailed presentation of the estimation of the survival function for right-censored data. We have stated in Section 2.6 that the Kaplan-Meier estimator is obtained by the formula $\hat{S}(t) = \prod_{j:\tilde{T}_j \leq t} \frac{n_j-d_j}{n_j}$. Since the estimator has a product term, the Kaplan-Meier estimator is also called as the *product-limit estimator*. We will briefly discuss a heuritic justification of the form. Since the publication of the original paper, it has seen over sixty-thousand citations. It is no surprise that the paper is very relevant and it will seemingly continue to do so in the future

Survival Analysis

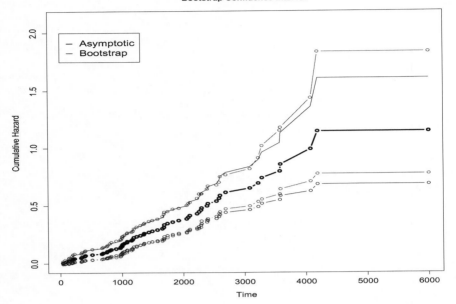

FIGURE 3.4: Bootstrap Confidence Intervals for the Nelson-Aalen Estimator

too. The work is also considered as one of the twenty important contributions in the subject of the twentieth century.

The estimator for the survival function can be derived in two ways. First, we follow the original pathbreaking approach. In the second, we exploit the relationship $S(t) = \exp(-H(t))$ and plug in the Nelson-Aalen estimator $\hat{H}(t)$ to obtain $\hat{S}(t) = \exp(-\hat{H}(t))$. We will first look at the heuristic explanation of the product-limit estimator.

We have the observed data in the form of $\tilde{T}_1 \leq \tilde{T}_2 \leq \ldots \leq \tilde{T}_n \leq \ldots$. This data consists of a mix of complete and censored observations. Consider the time interval $[0, t]$. Now, let $\breve{T}_1 \leq \breve{T}_2 \leq \ldots \leq \breve{T}_k$ be the event observed times in the interval $[0, t]$. We continue to assume that there are no ties. Partition the interval $[0, t]$ as $0 = t_0 < t_1 < \ldots < t_k = t$. Then, by the definition of the conditional probability, we have

$$S(t) = \prod_{j=1}^{k} S(t_j | t_{j-1}).$$

If an event occurs in the interval $(t_{j-1}, t_j]$, the estimate of $S(t_j | t_{j-1})$ is then $1 - 1/Y(t_{j-1})$. Further, if no event occurs in the interval, then the estimate is 1. Hence

$$\hat{S}(t) = \prod_{T_j \le t} \left\{ 1 - \frac{1}{Y(T_j)} \right\}. \tag{3.6}$$

The estimator $\hat{S}(t)$ is the famous Kaplan-Meier estimator.

The Product-integral Representation. In the continuous case, the cumulative hazard function can be written as

$$H(t) = -\int_0^t \frac{dS(u)}{S(u-)}.$$

and we can also write the hazard rate as $h(t) = -(1/S(t))dS(t)$. Consider the time interval $[0, t]$ and the partition $0 = t_0 < t_1 < t_2 < \ldots < t_k = t$. We now look at the finite product:

$$\prod_{j=1}^k \{1 - (H(t_j) - H(t_{j-1}))\}.$$

Increasing the number of intervals k while uniformly allowing the length of the interval to decrease to zero, the above product approaches a limit. The limit is called as *product-integral* and we denote it by $\prod_{s \le t} \{1 - dH(s)\}$, that is,

$$\prod_{s \le t} \{1 - dH(s)\} = \lim \prod_{j=1}^k \{1 - (H(t_j) - H(t_{j-1}))\}. \tag{3.7}$$

Consider the conditional probability $S(t_j | t_{j-1})$:

$$
\begin{aligned}
S(t_j | t_{j-1}) &= \frac{S(t_j)}{S(t_{j-1})} \\
&= 1 + \frac{S(t_j) - S(t_{j-1})}{S(t_j)} \\
&\approx 1 - (H(t_j) - H(t_{j-1}))
\end{aligned}
$$

Substituting for the conditional probability in the relation $S(t) = \prod_{j=1}^k S(t_j | t_{j-1})$, and taking the limits while increasing k as the length of the intervals uniformly approaches zero, we obtain

$$
\begin{aligned}
S(t) = \lim S(t) &= \lim \prod_{j=1}^k S(t_j | t_{j-1}) \\
&= \lim \prod_{j=1}^k \{1 - (H(t_j) - H(t_{j-1}))\} \\
&= \prod_{s \le t} \{1 - dH(s)\}. \tag{3.8}
\end{aligned}
$$

The survival function $S(t)$ as a product-integral of the cumulative hazard function also helps in establishing the relationship between the Kaplan-Meier estimator and the Nelson-Aalen estimator, again ABG[2]. Note that the Nelson-Aalen estimator can be written as $\hat{H}(t) = \sum_{T_j \leq t} \Delta \hat{H}(T_j)$. Plugging this in Equation 3.8, we obtain

$$\hat{S}(t) = \prod_{s \leq t} \{1 - d\hat{H}(s)\} = \prod_{T_j \leq t} \{1 - \Delta \hat{H}(T_j)\}. \tag{3.9}$$

We will define a quantity for $S(t)$ analogous to $H(t)$ in the form of $H^*(t)$ with

$$S^*(t) = \prod_{s \leq t} \{1 - dH^*(s)\}.$$

To establish the statistical properties of the Kaplan-Meier estimator, we need the following important relationship:

$$\frac{\hat{S}(t)}{S^*(t)} - 1 = -\int_0^t \frac{\hat{S}(s-)}{S^*(s)} d(\hat{H} - H^*)(s). \tag{3.10}$$

The integrand $\frac{\hat{S}(s-)}{S^*(s)}$ is a predictable process and its properties can be established based on the properties of the martingale $\hat{H} - H^*$. Clearly, the expression 3.10 represents a zero-mean martingale, and taking the expectations gives us $E\{\frac{\hat{S}(t)}{S^*(t)} - 1\} = 0$ and after simple algebra, we get $E\hat{S}(t) = S^*(t)$ which proves that the $\hat{S}(t)$ is an unbiased estimator for $S^*(t)$.

Using the asymptotic approximations of $\hat{S}(s-)/S^*(s) \approx 1$, $S^*(s) \approx S(s)$, and $H^*(s) \approx H(s)$ in Equation 3.10, we obtain

$$\frac{\hat{S}(t)}{S(t)} - 1 \approx -\int_0^t d(\hat{H} - H)(s),$$

$$\hat{S}(t) - S(t) \approx -S(t)(\hat{H}(t) - H(t)),$$

taking square of expectations to obtain variance

$$\mathbf{Var}(\hat{S}(t)) \approx S(t)^2 \mathbf{Var}(\hat{A}(t)),$$

using the estimates

$$\hat{\mathbf{Var}}(\hat{S}(t)) \approx \hat{S}(t)^2 \hat{\sigma}^2(t) = \hat{S}(t)^2 \sum_{T_j \leq t} \frac{1}{Y(T_j)^2}.$$

The Greenwoods formula for the variance of the Kaplan-Meier estimator is given by

$$\tilde{\mathbf{Var}}(\hat{S}(t)) = \hat{S}(t)^2 \sum_{T_j \leq t} \frac{1}{Y(T_j)\{Y(T_j) - 1\}}.$$

The asymptotic normality of the Kaplan-Meier estimator follows from the corresponding property of the Nelson-Aalen estimator. The $100(1-\alpha)\%$ confidence interval for $S(t)$ is given by

$$\hat{S}(t) \pm z_{1-\alpha/2}\hat{\text{Var}}(\hat{S}(t)). \qquad (3.11)$$

We will next apply these results on the PBC dataset.

Example 8 *Kaplan-Meier Estimators for the PBC Data.* *The Kaplan-Meier estimator was obtained earlier too, not used though. We will redo the calculations with the R software with a small difference. The formula in the* `survfit` *function is modified to* `Surv_PBC~trt`*. In Example 5, we used* 1 *instead of* `trt`*. What does the difference mean? Basically, we can create multiple Kaplan-Meier survival functions for the treatment levels. The Kaplan-Meier estimators are obtained first for the treatments and then at the gender level. The next code block accomplishes the estimates, and we also plot the Kaplan-Meier survival curves.*

```
> par(mfrow=c(1,2))
> Surv_PBC <- Surv(pbc$time,pbc$status==2)
> KM_TRT <- survfit(Surv_PBC~trt,data=pbc)
> print(KM_TRT)
Call: survfit(formula = Surv_PBC ~ trt, data = pbc)

   106 observations deleted due to missingness
        n events median 0.95LCL 0.95UCL
trt=1 158     65   3282    2583      NA
trt=2 154     60   3428    3090      NA
> plot(KM_TRT,col=c("red","blue"),conf.int = TRUE,
+       xlab="Days",ylab="S(t)",ylim=c(0,1.05))
> legend(0,0.6,c("D-penicillamine","Placebo"),pch="_",
+         col=c("red","blue"))
> title("PBC Survival Curves by Treatment")
> KM_Sex <- survfit(Surv_PBC~sex,data=pbc)
> print(KM_Sex)
Call: survfit(formula = Surv_PBC ~ sex, data = pbc)

        n events median 0.95LCL 0.95UCL
sex=m  44     24   2386    1536      NA
sex=f 374    137   3445    3170      NA
> plot(KM_Sex,col=c("red","blue"),conf.int = TRUE,
+       xlab="Days",ylab="S(t)",ylim=c(0,1.05))
> legend(0,0.6,c("Male","fEMALE"),pch="_",
+         col=c("red","blue"))
> title("PBC Survival Curves by Gender")
> dev.off()
null device
          1
```

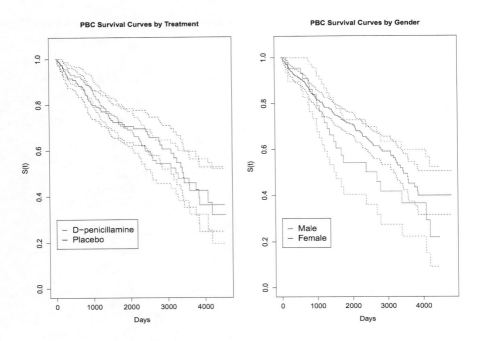

FIGURE 3.5: The Kaplan-Meier estimators

The result of the code block is Figure 3.5. In the left part of the figure, captioned by "PBC Survival Curves by Treatment", we can see that the survival curves of the treatment and placebo are overlapping. With the confidence intervals overlapping at almost every time point, the difference between the survival curves does not appear significantly different. Between 1500 and 3700 times, the probability of survival is higher for the placebo.

The difference between the survival curves in the gender is more pronounced. The survival probability all through is much less for males compared with females.

□

The mean and median survival times is important and we will take them next.

3.2.3 Mean and Median Survival Times

The survival curves and the cumulative hazard plots do not provide a single criterion of evaluation and the choice between two treatments might become subjective in certain clinical trials. In this context, if we are able to determine the average survival times, such as the mean and the median for the treatments, the decision making might become easier. Fortunately, we do not need to derive explicit methods to accomplish the tasks and the estimators of the aggregates will be based on the Kaplan-Meier estimator.

Median is a particular example of a percentile. It is the $p = 0.5$ percentile, and we consider the problem of finding the p-th percentile ξ_p of the cumulative distribution function $F(t)$. Note that even if ξ_p is uniquely determined by the relationship $1 - p = S(\xi_p)$, the value of $1 - p$ is not exactly attained because the estimator of the survival function S is a step-function. Hence, the estimate $\hat{\xi}_p$ is chosen as the smallest/earliest time at which $\hat{S}(t) \leq 1 - p$.

A well known result for a nonnegative random variable T with distribution function $F = 1 - S$ is that $E(T) = \int_0^\infty S(s)ds$. It often happens in the clinical trials that the largest survival time is censored because of the time constraint and we need to derive the estimate of the mean based on the data observed in the time interval $[0, t]$. We will call this estimate as the expected survival time restricted to the time interval $[0, t]$, and denote it by $\hat{\mu}_t = \int_0^t S(s)ds$. The estimate is provided by

$$\hat{\mu}_t = \int_0^t \hat{S}(s)ds,$$

where $\hat{S}(t)$ is the Kaplan-Meier estimator. An estimate of the variance of the mean restricted over the interval $[0, t]$ is given by

$$\hat{\text{Var}}(\hat{\mu}_t) = \sum_{T_j \leq t} \frac{(\hat{\mu}_t - \hat{\mu}_{T_j})^2}{Y(T_j)^2}.$$

We will continue illustrating these developments with the PBC dataset.

Example 9 *Obtaining the Mean and Median Survival Times for the **PBC Data**. The attentive reader would have already registered that we do not need to do anything extra for obtaining the median estimate because it pops up as a side-calculation on using the* survfit *function, as seen in Example 8. However, we will obtain the median survival times by treatment levels next and follow it by combination of the treatment with the gender.*

The computations related with the restricted mean time are performed in conjunction with the print *function. While passing a* survfit *object to the* print *function, the option* print.rmean = TRUE *must be included for the module to return the restricted mean. The default setting is to calculate the restricted mean time until the largest observed failure time. When we are interested in obtaining the restricted mean time at other times before the largest failure time, a further option needs to be specified in* rmean=time. *The formula can be extended for different levels of the categorical variables, such as treatment and gender. The following block computes the median and restricted mean times for multiple combinations.*

```
> # The Median and Mean Survival Times
> survfit(DSurv~1)
Call: survfit(formula = DSurv ~ 1)
```

```
       n   events   median 0.95LCL 0.95UCL
   158        65     3282     2583      NA
> pbcSurv <- Surv(pbc$time,pbc$status==2)
> survfit(pbcSurv~trt,data=pbc)
Call: survfit(formula = pbcSurv ~ trt, data = pbc)

   106 observations deleted due to missingness
         n events median 0.95LCL 0.95UCL
trt=1 158      65   3282     2583      NA
trt=2 154      60   3428     3090      NA
> survfit(pbcSurv~trt+sex,data=pbc)
Call: survfit(formula = pbcSurv ~ trt + sex, data = pbc)

   106 observations deleted due to missingness
               n events median 0.95LCL 0.95UCL
trt=1, sex=m  21     14   1682    1152      NA
trt=1, sex=f 137     51   3574    2598      NA
trt=2, sex=m  15      8   2796    1217      NA
trt=2, sex=f 139     52   3428    3170      NA
> print(survfit(DSurv~1),print.rmean = TRUE)
Call: survfit(formula = DSurv ~ 1)

          n      events    *rmean *se(rmean)    median
        158          65      2949        142      3282
     0.95LCL     0.95UCL
        2583          NA
     * restricted mean with upper limit =   4556
> print(survfit(DSurv~1),print.rmean = TRUE,rmean=3000)
Call: survfit(formula = DSurv ~ 1)

          n      events    *rmean *se(rmean)    median
      158.0        65.0    2289.5       78.1    3282.0
     0.95LCL     0.95UCL
      2583.0          NA
     * restricted mean with upper limit =   3000
> print(survfit(pbcSurv~trt,data=pbc),print.rmean = TRUE)
Call: survfit(formula = pbcSurv ~ trt, data = pbc)

   106 observations deleted due to missingness
         n events *rmean *se(rmean) median 0.95LCL
trt=1 158      65   2944        141   3282    2583
trt=2 154      60   2997        145   3428    3090
        0.95UCL
trt=1       NA
```

```
trt=2       NA
    * restricted mean with upper limit =   4540
> print(survfit(pbcSurv~trt+sex,data=pbc),print.rmean = TRUE)
Call: survfit(formula = pbcSurv ~ trt + sex, data = pbc)

    106 observations deleted due to missingness
                n events *rmean *se(rmean) median 0.95LCL
trt=1, sex=m  21     14   2283        348   1682    1152
trt=1, sex=f 137     51   3052        151   3574    2598
trt=2, sex=m  15      8   2730        445   2796    1217
trt=2, sex=f 139     52   3002        151   3428    3170
                0.95UCL
trt=1, sex=m        NA
trt=1, sex=f        NA
trt=2, sex=m        NA
trt=2, sex=f        NA
    * restricted mean with upper limit =   4491
```

The median survival time for the D-penicillamine is 3428 days and it is greater than that for the placebo at 3282 days by a margin of 146 days. However, the lower limit of the 95% confidence interval for the drug is 3090 days and hence, we can not say that the treatment is significantly better. The reader is asked to interpret the other results.

□

3.3 Smoothing the Hazard Rate

Mortality rate, the instantaneous failure rate, or the hazard rate, is of interest for practitioners. The cumulative hazard plot does not help in determining the risk at an exact time. In short, we need estimate of the hazard rate $h(t)$ at the time $t \in [0, \tau]$. The problem of directly estimating the hazard rate is as complex as that of estimating the density function, see Chapter 8 of Tattar, et al. (2016). The estimate of the hazard rate at time t as the difference of the Nelson-Aalen estimator $\hat{H}(t) - \hat{H}(t-)$ would be a futile exercise because it would be zero at all points except the failure times. Nevertheless, the Nelson-Aalen estimator can be used at the base and nonparametric smoothing techniques based on kernel can be used to estimate the hazard rate.

The kernels play an important role in smoothing. A kernel function $K(s)$ is a bounded nonnegative function defined over the interval $[-1, 1]$ and it satisfies $\int_{-1}^{1} K(s)ds = 1$. The three most widely useful kernel functions are given below:

- *The Uniform Kernel*: $K(s) = 1/2, -1 \le s \le 1$.

- *The Epanechnikov Kernel*: $K(s) = 3(1 - s^2)/4$.

- *The Biweight Kernel*: $K(s) = 15(1 - s^2)^2/16$.

The estimate of the hazard rate $h(t)$ at time t is obtained by the weighted average of the Nelson-Aalen increments $\Delta \hat{H}(T_j)$ over the interval $[t - b, t + b]$ by the following:

$$\hat{h}(t) = \frac{1}{b} \sum_{T_j \leq t} K \left(\frac{t - T_j}{b} \right) \Delta \hat{H}(T_j). \tag{3.12}$$

Here b is the bandwidth. The choice of kernel and the bandwidth are open end problems and as such there is no theoretical guidance over the specifications. It is better to experiment with different kernels and bandwidth combinations and check which plot appears more smooth while keeping an eye on the variance of the estimate of $\hat{h}(t)$.

The computations related to the kernel smoothing can get messy and we have an easy deployment in the R package `muhaz`. Using the same named function from the package, we can obtain the estimates of the hazard rate.

Example 10 *Smoothing the Hazard Rate for the PBC Data. Using the `muhaz` function with the option of `kern="rectangle"` and `bw.grid=500` and `bw.method="global"`, we smoothen the Nelson-Aalen estimator to produce the estimate of the hazard rate. We augment the plot of the rectangle smoothing kernel with Epanechnikov kernel and output Figure 3.6.*

```
> haz_rec <- muhaz(time=pbc.D$time,delta=pbc.D$status==2,
+                   bw.grid=500,kern="rectangle",
+                   bw.method="global", b.cor="none")
> pdf("../Output/Smooth_Hazard_Rate_PBC_D_Penicillamine.pdf",
+     height=10,width=13)
> plot(haz_rec)
> haz_epan <- muhaz(time=pbc.D$time,delta=pbc.D$status==2,
+                   bw.grid=500,kern="epanechnikov",
+                   bw.method="global", b.cor="none")
> points(haz_epan$est.grid,haz_epan$haz.est,"l",col="red")
> legend(0,4e-04,c("Rectangle","Epanechnikov"),pch="_",
+        col=c("black","red"))
> title("Smoothing the Hazard Rate")
> dev.off()
null device
          1
```

The hazard rate plot based on the rectangular kernel has lot of sharp edges. The Epanechnikov hazard rate plot appears more smooth and thereby for this example, we prefer it over the rectangular/uniform smoothing.

□

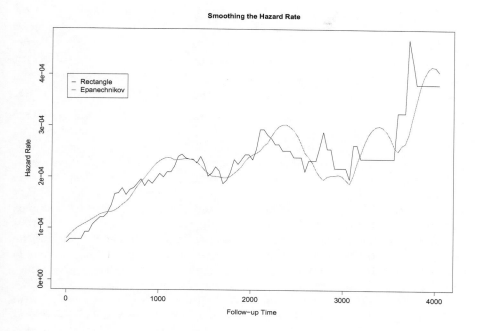

FIGURE 3.6: Smoothing the Hazard Rate

Smoothing techniques are important and it comes more handy for acturi-ans. The choice of the bandwidth might be validated using the bootstrap technique.

3.4 Estimation in Nonhomogeneous Markov Processes

Multi-state models (MSM) were introduced in Section 2.7. We saw two examples of competing risks model and health-related quality of life where the multi-state models are useful. The digraphs in Figures 2.5 and 2.6 show the possible transitions among the different states of the Markov process. Given the data that is subject to censoring, our goal is to determine the probabilities of transition among the various states. Here, the data will consist of the path of the patient traversing over the states. In this section, we will undertake the task of estimating the transition probability matrix (TPM) of the non-homogeneous Markov process. The notation from Chapter 2 will be revisited first.

The state function $X(t)$ will identify the state of the patient at time $t, t \in$ $\mathbb{R}^+$ with finite state space $\Gamma = \{1, 2, \ldots, k, 0\}$. The TPM over the interval $(s, t]$, is $\mathbf{P}(s, t) = \{P_{jj'}(s, t)\}$. For the state space Γ, we define the *transition rate* from state j to state j' by $h_{jj'}(t)$. The *cumulative intensity matrix* is given by $\mathbf{H} = \{H_{jj'}\}$, where $H_{jj'}(t) = \int_0^t h_{jj'}(s)ds$, and the TPM $\mathbf{P}(s, t)$ and $\mathbf{H(t)} = \{H_{jj'(t)}\}$ are related by $\mathbf{P}(s, t) = \prod_{(s,t]} (\mathbf{I} + d\mathbf{H}(u))$.

The time-to-event studies seen earlier are a particular case of the MSM with only two states -alive and dead. Here, we have transitions possible from multiple states before the patient is sucked into the absorbing state. In the earlier cases, the *counting process* $N(t)$ denoted counting the number of events occurring by time t. Now, we need to extend the counting process notation to register the number of transitions between different pairs of states. Let $\mathbf{N}(t) = \{N_{jj'}(t)\}$ denote the matrix-valued counting process with $N_{jj'}(t)$ counting the number of transitions starting in state j to state j' by time t. As before, we will assume that we have n subjects in the study. Further, let $Y_j(t)$ denote the number of subjects at-risk in state j at time t, $j \in \Gamma$, and $J_j(t) = I(Y_j(t) > 0)$. The estimator of the TPM $\mathbf{P}(s, t)$ is obtained by first deriving an estimator of the matrix of cumulative hazard functions $\mathbf{H(t)} = \{H_{jj'}(t)\}$. Aalen and Johansen (1978)[3] developed estimation of the TPM first and hence the estimator is also widely known as the *Aalen-Johansen estimator*. An estimator of the transition rate function, $H_{jj'}(t)$ is obtained as the Nelson-Aalen estimator:

$$\hat{H}_{jj'}(t) = \int_0^t \frac{dN_{jj'}(s)}{Y_j(s)}.$$

Let $\hat{\mathbf{H}}(t) = \{\hat{H}_{jj'}(t)\}$ denote the $(k + 1) \times (k + 1)$ matrix of the Nelson-Aalen estimators between pairs of states of the Markov process. The Aalen-Johansen estimator is then given by

$$\hat{\mathbf{P}}(s, t) = \prod_{(s,t]} \{\mathbf{I} + d\hat{\mathbf{H}}(u)\} = \prod_{s < T_j \leq t} \{\mathbf{I} + \Delta\hat{\mathbf{H}}(T_j)\}. \tag{3.13}$$

Now, for the state pair $\{j, j'\}$, define $H_{jj'}^*(t) = \int_0^t J_j(s)dH_{jj'}(s)$ and let $\mathbf{H}^*(t) = \{H_{jj'}^*(t)$. The TPM associated with $\mathbf{H}^*(t)$ is then $\mathbf{P}^*(s, t) = \prod_{(s,t]} \{\mathbf{I} + \mathbf{H}^*(u)\}$. Duhamel's equation helps us to setup the relationship that gives the martingale property of the Aalen-Johansen estimator, refer ABG for details, and toward that we have the following:

$$\hat{\mathbf{P}}(s, t)\mathbf{P}^*(s, t)^{-1} - \mathbf{I} = \int_{(s,t]} \hat{\mathbf{P}}(s, u-)d\left(\hat{\mathbf{H}} - \mathbf{H}^*\right)(u)\mathbf{P}^*(s, u)^{-1}. \tag{3.14}$$

It can be easily proved that $\hat{\mathbf{P}}(s, t)\mathbf{P}^*(s, t)^{-1} - \mathbf{I}$ is a matrix of zero-mean martingale and thereby $\hat{\mathbf{P}}(s, t)$ will be an unbiased estimator of $\mathbf{P}^*(s, t)$. For more details, the reader can refer ABG and ABGK.

In the next example, we will illustrate the computation of TPM estimator for the necropsy data.

Example 11 *Aalen-Johansen Estimator for the Mice Necropsy Data. ABGK's Example I.3.8. details this laboratory experiment. Here, we have two types of mice—the conventional mice (95) and the germ-free mice (82). The mice are exposed to radiation dose of 300 rad when they are 5–6 weeks in age. After a mouse dies, a necropsy test is performed and the cause is identified as thymic lymphoma, recticulum cell sarcoma, or other causes. The purpose of the study is to understand the cause of the deaths under the different setups. This is an example of competing risks model.*

The data is available in file Necropsy_Mice.csv. *Here, we have four states with "Alive" as one state and the other three are the causes of death. The* etm *package is used to obtain the Aalen-Johansen estimator. We need to import the data first, mark the cause of death, note the time of the death, and specify the cause of death. The* etm *function is then used to do the necessary computations.*

```
> # Estimation in Nonhomogeneous Markov Processes
> # Analysis of Necropsy Mice Data
> mice <- read.csv("../Data/Necropsy_Mice.csv",header=TRUE,
+                  stringsAsFactors = FALSE)
> str(mice)
'data.frame': 177 obs. of  6 variables:
 $ time      : int  159 189 191 198 200 207 220 235 245 250 ...
 $ Cause     : chr  "Thymic Lymphoma" ... "Thymic Lymphoma" ...
 $ Mice.Type : chr  "Conventional" "Conventional" ...
 $ from      : chr  "Alive" "Alive" "Alive" "Alive" ...
 $ State     : int  1 1 1 1 1 1 1 1 1 1 ...
 $ to        : chr  "Thymic" "Thymic" "Thymic" "Thymic" ...
> mice$to <- as.character(mice$to)
> mice$id <- paste0("Mice_",1:nrow(mice))
> SN <- c("Alive","Thymic","Recticulum","Other") # State Names
> NM_Trans <- matrix(FALSE,nrow=4,ncol=4)
> rownames(NM_Trans) <- SN
> colnames(NM_Trans) <- SN
> NM_Trans[1,2:4] <- TRUE
> NM_Trans
           Alive Thymic Recticulum Other
Alive      FALSE  TRUE        TRUE  TRUE
Thymic     FALSE FALSE       FALSE FALSE
Recticulum FALSE FALSE       FALSE FALSE
Other      FALSE FALSE       FALSE FALSE
> mice_Con <- mice[mice$Mice.Type=="Conventional",]
> Mice_AJ_Con <- etm(mice_Con,state.names = SN,tra=NM_Trans,
+                    s=1,cens.name = NULL)
> print(Mice_AJ_Con)
```

```
Multistate model with 1 transient state(s)
 and 3 absorbing state(s)

Possible transitions:
  from           to
 Alive       Thymic
 Alive Recticulum
 Alive       Other

Estimate of P(1, 763)
           Alive Thymic Recticulum  Other
Alive          0 0.2316        0.4 0.3684
Thymic         0 1.0000        0.0 0.0000
Recticulum     0 0.0000        1.0 0.0000
Other          0 0.0000        0.0 1.0000
```

```
> pdf("../Output/Mice_Necropsy_Conventional_Aalen_Johansen.pdf",
+      height=12,width=12)
> xyplot(Mice_AJ_Con,
+        tr.choice = c("Alive Alive","Alive Thymic",
+            "Alive Recticulum","Alive Other"),
+        layout=c(2,2)
+        )
> dev.off()
null device
          1
```

The output of the code block is the transition probabilities and it is seen in Figure 3.7. The transition probability curves are accompanied by the confidence intervals. From the transition probability plots of 'Alive Recticulum', 'Alive Other', and 'Alive Thymic', we can conclude that after 500 days the probability of a living mice dying because of recticulum cell sarcoma is higher than dyeing due to the other two causes. However, in the initial period, the deaths are more often due to thymic lymphoma.

□

3.5   Exercises

Exercise 3.1 *Comparing the D-penicillamine treatment with placebo. In continuation of Example 5, obtain the confidence intervals for the placebo too and update Figure 3.3. Report your conclusion.*   □

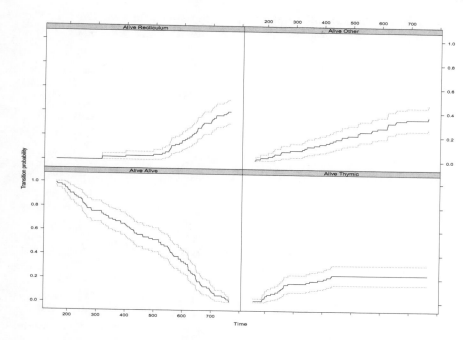

FIGURE 3.7: The Aalen-Johansen Estimator for the Mice Necropsy Data

Exercise 3.2 *Simulate $n = 1000$ observations from an Weibull distribution with failure rate $1/100$ and shape parameter 1.5. Assume that censored times follow Weibull with rate $1/50$ and same shape parameter 1.5. Fit the Nelson-Aalen cumulative hazard function and compare it with the true cumulative hazard function.*

☐

Exercise 3.3 ***Comparing Bootstrap Confidence Intervals of the D-penicillamine treatment with placebo****. Continue Example 7 further and obtain the bootstrap confidence intervals for the placebo too and draw your conclusion.*

☐

Exercise 3.4 *Extend estimation of the hazard rate in Example 10 for the uniform and biweight kernel.*

Exercise 3.5 ***Estimate TPM for Germ-free Mice****. Compute the Aalen-Johansen estimator for the germ-free mice and compare it with the conventional ones.*

☐

Exercise 3.6 *Sojourn/waiting time distributions are important in the analysis of multi-state models, especially for progressive states model. Here, we are more interested in the probability of finding a patient in a particular state.*

The R package `msSurv` *helps in obtaining these probabilities. The user should explore the package to obtain the probabilities of fitting the sojourn time distributions.*

□

3.6   Más Lejos Temas

The asymptotic properties of the Kaplan-Meier estimator were established in Kaplan and Meier (1958)[61] and the stochastic calculus proof in Aalen (1975)[4] and the point process derivation in Karr (1986)[62]. The product-limit estimator also has many other useful statistical properties. For instance, it is a self-consistent estimator, see Miller (1981)[82]. Using the concept of integrals and using the Kaplan-Meier integrals, Stute (2003)[108] establishes the strong law of large numbers, or almost-sure convergence.

Empirical risk processes have rich applications in the broader area of stochastic processes, and a benchmark reference in the context of their applications to statistics can be found in Shorack and Wellner (1986)[106]. Statistical inference and properties of the Kaplan-Meier estimator as an empirical process is delved deeper in Kosorok (2008)[65]. The bootstrap approach to Kaplan-Meier estimator is worked out in Efron (1981)[40] and Akritas (1986)[6].

Smoothing techniques took off with the advent of computational power. The choice of the smoothing kernels and the bandwidth are as such open end problems. A more satisfactory solution might be acceptable under the bootstrap method. Section 6.2 of Klein and Moeschberger (2003)[64] discusses the details related to the smoothing techniques.

We used the nonhomogeneous Markov process for multi-state model. Huzurbazar (2005)[56] considers the flowgraph approach to multi-state models, and the monograph has a lot of applications in reliability theory. Section 3.4 gives us the Aalen-Johansen estimator for estimation of the transition probability matrix associated with the nonhomogeneous Markov process. The bootstrap method for the Aalen-Johansen estimator is derived extensively in James (1998)[58]. Willekens (2014)[119] and Beyersmann, et al. (2011)[11] comprehensively carried out analysis of multi-state models using the R software. Brostrom (2018)[20] is another recent textbook on survival data and R.

Chapter 4

Inference—Statistical Tests

4.1 Introduction

Chapter 2 gave us techniques of fitting parametric distributions to the lifetime data and we found several families of distributions that were useful towards that. In Chapter 3, nonparametric estimation of survival functions were developed extensively. We had developed the smoothing techniques for understanding the hazard rate. When comparing two or more treatments, the survival curves and the cumulative hazard plots provide guidance to assess efficacy of a drug/treatment. While similarity can be concurred if the curves are much similar, graphical evidence can not be used to conclude that the efficacies are significantly different. Thus, we need appropriate statistical tests to conclude whether the efficacies are significantly different. This chapter will begin with parametric tests.

In the case of parametric distributions, the survival function is completely determined by the parameters, and thus test for equality of survival curves can be carried out by hypothesis testing related to the unknown parameters. Two tests are developed for testing the equality of two exponential distributions—chi-square test and the F-test. The tests are based on the likelihood ratio test. Since we do not have an in-built R function for carrying out the hypothesis testing, we create two new functions that accomplish the illustration.

Nonparametric methods are of paramount importance as seen in Chapter 3. We need nonparametric tests to continue the work of related estimation problems. One-sample and k-sample tests are then developed and deployed to carryout the suitable tests. All tests are illustrated with practical dataset.

4.2 Parametric Tests

We introduced a large class of families of probability distributions in Chapter 2—exponential, gamma, Weibul, generalized distributions, and so forth. Fitting distributions by estimating the parameters using the MLE technique had been performed earlier. Given lifetimes of n subjects, some possibly cen-

DOI: 10.1201/9781003306979-4

sored, and the unknown failure time distribution, we first express the likelihood function for right-censored data in terms of the counting process and the at-risk process. The likelihood function plays an important role in carrying out hypothesis testing since many important class of tests are based on the likelihood ratios.

Let us denote the lifetimes of n individuals by $T_1, T_2, \ldots, T_n$, and their respective censoring variables with $C_1, C_2, \ldots, C_n$. We observe $(\tilde{T}_i, \delta_i), i = 1, 2, \ldots, n$, with $\tilde{T}_i = T_i \wedge C_i$ and $\delta_i = I(\tilde{T}_i = T_i)$. The aggregated counting process is $N(t) = \sum_{i=1}^{n} N_i(t)$ with $N_i(t)$ being 0 if the event is observed for the i-th data point by time t. Further, let $Y(t) = \sum_{i=1}^{n} Y_i(t)$ be the at-risk process. The marked difference from the earlier chapters, and in this section, is that $F(t)$ and thereby $S(t)$ are parametric functions. It implies that the survival function and the hazard rate are expressed as $S(t, \theta)$ and $h(t, \theta)$. The likelihood contribution of the i-th patient is

$$L_i(\theta | \tilde{t}_i, \delta_i) = f(\tilde{t}_i, \theta)^{\delta_i} S(\tilde{t}_i, \theta)^{1-\delta_i} = h(\tilde{t}_i, \theta)^{\delta_i} \exp\left\{ -\int_0^{\tilde{t}_i} h(s, \theta) ds \right\}.$$

The likelihood function for a random sample of size n is therefore

$$L(\theta | \texttt{data}) = \prod_{i=1}^{n} h(\tilde{t}_i, \theta)^{\delta_i} \exp\left\{ -\int_0^{\tilde{t}_i} h(s, \theta) ds \right\}.$$

Let τ denote the upper time limit of the study, or the largest observed failure time. Also, let $\Delta N_i(t) = N_i(t) - N_i(t-), h_i(t, \theta) = Y_i(t) h(t, \theta)$. Then, the likelihood function corresponding to the i-th observation is

$$L_i(\theta) = \left\{ \prod_{0 < t \leq \tau} h_i(t, \theta)^{\Delta N_i(t)} \right\} \exp\left\{ -\int_0^{\tau} h_i(t, \theta) dt \right\}.$$

Thus, the likelihood function for a sample of size n takes the form:

$$L(\theta) = \left\{ \prod_{i=1}^{n} \prod_{0 < t \leq \tau} h_i(t, \theta)^{\Delta N_i(t)} \right\} \exp\left\{ -\int_0^{\tau} \sum_{i=1}^{n} h_i(t, \theta) dt \right\}.$$

Now, consider the problem of comparing the effectiveness of two treatments, for instance, we had placebo and D-penicillamine in the Primary Biliary Cirrhosis dataset. Based on random samples of patients who have undergone the treatments, we would like to test if the efficacies are equal, or one of them is better over the other. We will introduce a new notation to denote the observations and parameters of the two populations under consideration. Let S_1 and S_2 denote the survival functions corresponding to the two treatments. We assume that the parameters governing the two distributions are respectively θ_1 and θ_2. We assume that each θ has p number of parameters. The statistical hypotheses testing problem is then testing the hypothesis $H_0 : S_1 = S_2$

against the alternative $H_1 : S_1 \neq S_2$ i.e., S_1 and S_2 differ at least at one t. The problem is also equivalent to testing $H_0 : \theta_1 = \theta_2$ against $H_1 : \theta_1 \neq \theta_2$.

Let n_l random observations of population $l, l = 1, 2$, be denoted by $T_{l,1}, T_{l,2}, \ldots, T_{l,n_l}, l = 1, 2$, and their respective censoring variables by $C_{l,1}, C_{l,2}, \ldots, C_{l,n_l}$. We then have $(\tilde{T}_{l,i}, \delta_{l,i}), i = 1, 2, \ldots, n_l$, as the observed data and the corresponding counting process and at-risk process are given by $N_l(t) = \sum_{i=1}^{n_l} N_{l,i}, Y_l(t) = \sum_{i=1}^{n_l} Y_{l,i}(t), l = 1, 2$. The l-th likelihood function corresponding to population l is then given by

$$L_l(\theta_l) = \left\{ \prod_{i=1}^{n_l} \prod_{0 < t \leq \tau} h_i(t, \theta_l)^{\Delta N_{l,i}(t)} \right\} \exp \left\{ -\int_0^\tau \sum_{i=1}^{n_l} h_{l,i}(t, \theta_l) dt \right\}, l = 1, 2.$$

The joint likelihood function for both the samples, by independence of the samples, is

$$L(\theta_1, \theta_2) = L(\theta_1) \times L(\theta_2).$$

It is generally convenient to work with the loglikelihood functions, and it will easily follow that $\log L(\theta_1, \theta_2) = \log L(\theta_1) + \log L(\theta_2)$.

Under the null hypothesis $H_0 : \theta_1 = \theta_2 = \theta$, say, the parameters of the two populations are equal and it is meaningful to combine the two samples and estimate the unknown parameter θ. Let us denote the ML estimates of the parameters by $\hat{\theta}_1, \hat{\theta}_2, \hat{\theta}$. The log likelihood function under the null hypothesis based on the $n_1 + n_2$ observations is denoted by $\log L(\theta)$.

The **Log-likelihood Ratio test statistic** is given by

$$LR = -2 \left(\log L(\hat{\theta}) - \log L(\hat{\theta}_1) - \log L(\hat{\theta}_2) \right).$$

It follows from large-sample theory that LR follows a chi-square distribution with p degrees of freedom. Hence, at significance level α, we reject the null hypothesis that the treatments are equally effective if $LR > \chi^2_{p,\alpha}$, where $\chi_{p,\alpha/2}$ is the $(1 - \alpha)$ quantile of a chi-square distribution with p degrees of freedom. We will next develop the likelihood ratio test statistic for the right-censored data from exponential distribution.

Example 12 *Testing for Equality of Exponential Curves. Consider two treatment groups in which the observations are right-censored. With respect to the description given in the section, we have $S_1(t) = \exp\{-t/\lambda_1\}$ and $S_2(t) = \exp\{-t/\lambda_2\}$. Let λ denote the common parameter. For simplicity, we assume that the first r_1 observations from population 1 are complete observations, that is, $\delta_{1,i} = 1, i = 1, \ldots, r_1$, and $\delta_{1,i} = 0, i = r_1 + 1, \ldots, n_1$. Similarly, we assume that the first r_2 observations are complete observations for population 2. It is then straightforward to obtain the MLEs of the parameters, and we state them here:*

$$\hat{\lambda}_l = \frac{\sum_{i=1}^{r_l} t_{l,i} + \sum_{i=r_l+1}^{n_l} t_{l,i}}{r_l}, l = 1, 2,$$

$$\hat{\lambda} = \frac{\sum_{l=1,2} \left\{ \sum_{i=1}^{r_l} t_{l,i} + \sum_{i=r_l+1}^{n_l} t_{l,i} \right\}}{r_1 + r_2}.$$

In terms of the preceding notations, we can write the parameters as follows,
$\theta_1 = \lambda_1, \theta_2 = \lambda_2, \theta = \lambda$. *Using the estimators $\hat{\lambda}_l, l = 1, 2$ and $\hat{\lambda}$, we next calculate the likelihoods:*

$$L(\lambda_l) \;=\; \frac{\exp\left\{-\left(\sum_{i=1}^{r_l} t_{l,i} + \sum_{i=r_l+1}^{n_l} t_{l,i}\right)/\hat{\lambda}_l\right\}}{\hat{\lambda}_l^{r_l}}, l = 1, 2,$$

$$L(\lambda) \;=\; \frac{\exp\left\{-\sum_{l=1,2}\left(\sum_{i=1}^{r_l} t_{l,i} + \sum_{i=r_l+1}^{n_l} t_{l,i}\right)\right\}}{\hat{\lambda}_1^{r_1}\hat{\lambda}_2^{r_2}}.$$

Essentially, we require the MLE's of the failure rate under both the populations first, and then combining the samples, we obtain the common failure rate. The estimates then drive the corresponding likelihood functions and enable calculation of the likelihood ratio test statistic. The function `Exp_Surv_LR_Test` *carries out the required computations and at a specified level of significane, the hypothesis test of equality of the exponential survival curves is performed.*

```
> Exp_Surv_LR_Test <- function(t1,d1,t2,d2,alpha){
+    lambda1 <- sum(t1)/sum(d1) #MLE of Lambda1
+    lambda2 <- sum(t2)/sum(d2) #MLE of Lambda2
+    lambdap <- (sum(t1)+sum(t2))/(sum(d1)+sum(d2)) #Pooled MLE
+    likeli_denom <- exp(-(sum(t1)/lambda1)-(sum(t2)/lambda2))/
+      ((lambda1^sum(d1))*(lambda2^sum(d2)))
+    likeli_p <- exp(-(sum(t1)+sum(t2))/lambdap)/
+      (lambdap^(sum(d1)+sum(d2)))
+    LR <- -2*log(likeli_p/likeli_denom)
+    print(c(LR,qchisq(1-alpha,1)))
+    if(LR>qchisq(1-alpha,1)){
+            print("Exponential Survival Curves are Different")}
+    if(LR<=qchisq(1-alpha,1)){
+            print("Not Enough Data Points to
+            Reject Equality of Survival Curves")}
+    }
```

For the remission data of two treatment groups, see page 248 of Lee and Wang (2003)[71], we apply the R test function `Exp_Surv_LR_Test`. *The calculated loglikelihood ratio test statis is lesser than the tabulated value at $\alpha = 0.05$ level of significance and hence we conclude that we do not have sufficient evidence to reject the equality of survival curves.*

```
> t1 <- c(23,16,18,20,24)
> d1 <- c(1,0,0,0,0)
> t2 <- c(15,18,19,19,20)
> d2 <- rep(1,5)
> Exp_Surv_LR_Test(t1,d1,t2,d2,alpha=0.05)
[1] 3.344374 3.841459
```

[1] "Not Enough Data Points to
+   Reject Equality of Survival Curves"

□

An alternative test procedure is based on the Cox's F-test. The procedure was first derived in Cox (1953)[31] and also detailed in Lee and Wang (2003)[71]. Here, we simply obtain $F = \hat{\lambda}_1/\hat{\lambda}_2$, where $\hat{\lambda}_l = \frac{\sum_{i=1}^{r_l} t_{l,i} + \sum_{i=r_l+1}^{n_l} t_{l,i}}{r_l}, l = 1, 2$. The Cox's F-test procedure is to reject the null hypothesis if $F > F_{2r_1, 2r_2, \alpha/2}$. The $100(1 - \alpha)$ confidence interval is provided by $\left[\hat{\lambda}_2/\hat{\lambda}_1 F_{2r_1, 2r_2, 1-\alpha/2}, \hat{\lambda}_2/\hat{\lambda}_1 F_{2r_1, 2r_2, \alpha/2} \right]$. We will illustrate these ideas with a practical dataset.

Example 13 *Testing for Equality of Exponential Curves Using Cox's F-Test. The data here is drawn from Example 10.2 on page 250 of Lee and Wang (2003)[71]. Here, thirty-six patients with glioblastoma multiforme were divided into two groups—an experimental group and the control group. The experimental group considered of surgery and chemotherapy treatment and this group consists of twenty-one patients. In the control group, the treatment consisted of surgery alone and it was given to fifteen patients. We need to test whether the survival times in the two treatment are equal or otherwise.*

We create a new function Cox_F_Test. *The R computations are straightforward and do not need further explanation.*

```
> # Cox's F-Test for Equality of Exponential Survival Curves
> Cox_F_Test <- function(t1,d1,t2,d2,alpha){
+    r1 <- sum(d1); r2 <- sum(d2)
+    t1bar <- sum(t1)/r1
+    t2bar <- sum(t2)/r2
+    UCL <- qf(1-alpha/2,2*r1,2*r2)*t2bar/t1bar
+    LCL <- qf(alpha/2,2*r1,2*r2)*t2bar/t1bar
+    print(c(LCL,UCL))
+    print(t2bar/t1bar)
+ }
> EG_t <- c(1,2,2,2,6,8,8,9,13,16,17,29,34,
+           2,9,13,22,25,36,43,45)
> EG_d <- rep(c(1,0),times=c(13,8))
> CG_t <- c(0,2,5,7,12,42,46,54,7,11,19,22,30,35,39)
> CG_d <- rep(c(1,0),times=c(8,7))
> Cox_F_Test(EG_t,EG_d,CG_t,CG_d,alpha=0.05)
[1] 0.6665018 4.0943320
[1] 1.572734
```

Since 1 is in the 95% confidence interval, the difference between the two treatments is not statistically significant. □

The extensions to other parametric distributions is possible with appropriate tweaks. We will now focus on the nonparametric tests.

4.3 One-sample Nonparametric Tests

In certain scenarios, we might have a benchmark failure rate or a specific survival model in mind that needs to be improved by the proposed treatment. In such cases, samples are obtained for the proposed treatment and we need to test whether it meets the current standards of efficacy or betters it. In short, the hypothesis of interest is $H_0 : S = S_0$, equivalently, in terms of the hazard rate, $H_0 : h = h_0$, where S_0, or h_0, is the benchmark survival function. We would like to build the test statistic based on the Nelson-Aalen estimator as developed in Chapter 3.

As per the earlier convention, we will denote the cumulative hazard function by $H(t) = \int_0^t h(s)ds$, the Nelson-Aalen estimator $\hat{H}(t) = \int_0^t J(s)/Y(s)dN(s)$, where $Y(t)$ and $N(t)$ are respectively the at-risk and counting process (of the events) while $J(t) = I(Y(t) > 0)$. The central idea of the test statistic is the comparison of the Nelson-Aalen increments with the expected number of events under the specified survival distribution. Weight functions provide a critical theme in comparison of the observed number of events with the hypothesized values. We denote the nonnegative weight function by $K(t)$. A lot of choices are available for the $K(t)$, and the popular choices are the at-risk process, that is, $K(t) = Y(t)$, square-root of the weight process, $K(t) = Y^{1/2}(t)$, and sometimes as a function of the survival function $K(t) = \hat{S}^\rho(t-), 0 \leq \rho \leq 1$.

Consider the function $H^*(t) = \int_0^t J(s)h(s)ds$ under the null hypothesis $H_0 : h = h_0$:

$$H_0^*(t) = \int_0^t J(s)h_0(s)ds.$$

It can be shown under the null hypothesis that the difference between the Nelson-Aalen integral and $H_0^*(t)$ in $\hat{H}(t) - H_0^*(t)$ is a local-square integrable martingale. The general test statistic based on the predictable and nonnegative weight function $K(t)$ is given by

$$Z(t) = \int_0^t K(s)d(\hat{H} - H_0^*)(s). \tag{4.1}$$

Note that the predictable variation process of $\hat{H}(t) - H_0^*(t)$ is $\left\langle \hat{H} - H_0^* \right\rangle (t) = \int_0^t J(s)h_0(s)/Y(s)ds$. Consequently, the predictable variation process of the test statistic is

$$\langle Z \rangle(t) = \int_0^t K^2(s) d\langle \hat{H} - H^* \rangle(s) = \int_0^t K^2(s) \frac{h_0(s)}{Y(s)} ds. \qquad (4.2)$$

Specification of the choice of the weight process allows us pick appropriate test statistics. For instance, if we choose $K(t) = Y(t)$, we obtain the log-rank test statistic, and in this case we have the following:

$$
\begin{aligned}
Z(t) &= \int_0^t Y(s) \left\{ \frac{J(s)}{Y(s)} dN(s) - J(s)h_0(s)ds \right\} = N(t) - E(t), \\
\langle Z \rangle(t) &= E(t),
\end{aligned}
$$

where $E(t) = \int_0^t h_0(s)Y(s)ds$ is the expected number of events under the null hypothesis.

For more on the one-sample test, one may refer to Section 1, Chapter 5 of ABGK. To perform the one-sample test in R, we need to first define the expected number of events at the jump points of the Nelson-Aalen integral. Using the option of `offset` and `survdiff` function, we can carry out the log-rank test.

Example 14 *Testing for Experimental Group Survival Function. We will continue with the study in Example 13. Assuming that the control group mean is the true mean and that the survival times follow an exponential distribution, we would like to test if the experimental group survival times also follow the same distribution. The following R program accomplishes the required task.*

```
> # One-sample Nonparametric Tests
> CG_lambda <- sum(CG_t)/sum(CG_d)
> Expect_EG_Times <- 1-pexp(EG_t,1/CG_lambda)
> survdiff(Surv(EG_t,EG_d)~offset(Expect_EG_Times))
Call:
survdiff(formula = Surv(EG_t, EG_d) ~ offset(Expect_EG_Times))

Observed Expected        Z        p
 13.0000   8.2659  -1.6466   0.0996
```

At 5% level of significance, we do not have enough evidence to reject the hypothesis that the survival times of the experimental group is the same as the control group. □

Note that the above technique is versatile and all that we require is the expected number of events at jump points of the Nelson-Aalen integral. This can be easily performed so long as we are able to compute the probabilities under $H_0 : h = h_0$.

4.4 k-sample Nonparametric Tests

We require extension of the one-sample test discussed in the preceding section to the more practical case of comparison of two or more treatments. We will call this as the k-sample test. The survival curves being compared need not be restricted to data from administering of treatments. It can be extended to cover the categorical variables such as gender, region, race, etc. The main idea remains the same as we proceed with comparison of multiple treatments. The k-sample tests will be again based on the Nelson-Aalen integrals. The cumulative hazard estimates will be obtained for each of the k-treatments first. Next, we combine all the observations as if they were obtained from a single source and obtain the pooled estimate of the cumulative hazard function. We will setup the framework now.

Let $S_1, S_2, \ldots, S_k$ denote the k survival curves as response to the k treatments. We assume that the survival functions are absolutely continuous. The results in the section continue to hold, with minor tweaks, for the general cases too. The hazard rates associated with the respective treatments are $h_1, h_2, \ldots, h_k$. The statistical hypotheses testing problem is then testing $H_0 : S_1 = S_2 = \ldots = S_k$ against the alternative $H_1 : S_l \neq S_{l'}, l \neq l' = 1, 2, \ldots, k$, for at least one pair of indexes. The hypotheses can be equivalently stated in terms of the hazard rates $H_0 : h_1 = h_2 = \ldots = h_k$ against $H_1 : h_l \neq h_{l'}, l \neq l' = 1, \ldots, k$, for at least one pair of indexes. Let n_l random observations of population $l, l = 1, 2, \ldots, k$, be denoted by $T_{l,1}, T_{l,2}, \ldots, T_{l,n_l}, l = 1, \ldots, k$, and their respective censoring variables be $C_{l,1}, C_{l,2}, \ldots, C_{l,n_l}$. Similarly, we have $(\tilde{T}_{l,i}, \delta_{l,i}), i = 1, 2, \ldots, n_l$, as the observed data, $N_l(t) = \sum_{i=1}^{n_l} N_{l,i}, Y_l(t) = \sum_{i=1}^{n_l} Y_{l,i}(t), l = 1, \ldots, k$, as the counting processes and the at-risk processes. The Nelson-Aalen estimator for the l-th population is

$$\hat{H}_l(t) = \int_0^t \frac{J_l(s)}{Y_l(s)} dN_l(s).$$

The common survival function can be denoted by S, that is, under the null hypothesis we have $H_0 : S_1 = \ldots = S_k = S$, and equivalently $H_0 : h_1 = \ldots = h_k$. Now, let $N_\bullet(t) = \sum_{l=1}^k N_l(t)$, $Y_\bullet(t) = \sum_{l=1}^k Y_l(t)$, and $J_\bullet(t) = I(Y_\bullet(t) > 0)$. The common cumulative hazard function is then estimated by

$$\hat{H}(t) = \int_0^t \frac{J_\bullet(s)}{Y_\bullet(s)} dN_\bullet(s).$$

To carry out the hypothesis testing, we compare the increments of $\hat{H}_l(t)$ with $\hat{H}(t)$ for values of t where $Y_l(t) > 0$. In this direction, we need the following:

$$\tilde{H}_l(t) = \int_0^t J_l(s) d\hat{H}(t) = \int_0^t \frac{J_l(s)}{Y_\bullet(s)} dN_\bullet(s).$$

Recall that the martingale equation associated with N_l is

$$\int_0^t \frac{J_l(s)}{Y_l(s)} dN_l(s) = \int_0^t J_l(s) h_l(s) ds + \int_0^t \frac{J_l(s)}{Y_l(s)} dM_l(s), l = 1, \ldots, k.$$

Thus, when the null hypothesis $H_0 : h_1 = \ldots = h_k$ holds true, we can obtain the difference between $\hat{H}_l(t)$ and $\tilde{H}_l(t)$ as follows:

$$\hat{H}_l(t) - \tilde{H}_l(t) = \int_0^t \frac{J_l(s)}{Y_l(s)} dM_l(s) - \int_0^t \frac{J_l(s)}{Y_\bullet(s)} dM_\bullet(s),$$

where $M_\bullet(t) = \sum_{l=1}^k M_l(s)$. It can be easily seen that under the null hypothesis $\hat{H}_l(t) - \tilde{H}_l(t)$ is a local square integrable martingale. We will next accumulate the differences through a weight function $K_l(t)$. As with the single-sample case, a host of options are available for the weight process. For simplicity, we will have a common weight function across the samples and allow the variation only through the at-risk process at time t, that is, we will choose the specification $K_l(t) = Y_l(t)K(t), l = 1, \ldots, k$. In addition to the requirement of $K(t)$ being a nonnegative, locally bounded predictable process, we will require it to depend only on $N_\bullet$ and $Y_\bullet$. The difference integral function is then defined in the following:

$$
\begin{aligned}
Z_l(t) &= \int_0^t K_l(s) d(\hat{H}_l - \tilde{H}_l)(s) \\
&= \int_0^t Y_l(s) K(s) d(\hat{H}_l - \tilde{H}_l)(s) \\
&= \int_0^t K(s) dN_l(s) - \int_0^t K(s) \frac{Y_l(s)}{Y_\bullet(s)} dN_\bullet(s), l = 1, \ldots, k. \quad (4.3)
\end{aligned}
$$

The reader can easily see that $\sum_{l=1}^k Z_l(t) = 0$. Using the difference $\hat{H}_l(t) - \tilde{H}_l(t)$ in terms of the martingales $M_l(t)$ and $M_\bullet(t)$ in the second step of Equation 4.3, we obtain

$$Z_l(t) = \int_0^t K(s) dM_l(s) - \int_0^t K(s) \frac{Y_l(s)}{Y_\bullet(s)} dM_\bullet(s).$$

Using the Kroneckar delta $\delta_{ll'} = I\{l = l'\}$, the above expression simplifies to

$$Z_l(t) = \sum_{l'=1}^k \int_0^t K(s) \left(\delta_{ll'} - \frac{Y_l(s)}{Y_\bullet(s)} \right) dM'_l(s). \quad (4.4)$$

With the integrand being a predictable process, it follows that $Z_l(t)$ is a local square integrable martingale, and hence the predictable covariation process between Z_l and Z_m is

$$\langle Z_l, Z_m \rangle(t) = \int_0^t K^2(s) \frac{Y_l(s)}{Y_\bullet(s)} \left(\delta_{lm} - \frac{Y_m(s)}{Y_\bullet(s)} \right) h(s) Y_\bullet(s) ds, \quad (4.5)$$

and an unbiased estimator of the covariation process is given by

$$\hat{\sigma}_{lm}(t) = \int_0^t K^2(s) \frac{Y_l(s)}{Y_\bullet(s)} \left(\delta_{lm} - \frac{Y_m(s)}{Y_\bullet(s)} \right) dN_\bullet(s). \tag{4.6}$$

Thanks to the constraint $\sum_{l=1}^k Z_l(t) = 0$, we can drop one of the elements $Z_1(t), Z_2(t), \ldots, Z_k(t)$, and we will leave out $Z_k(t)$. Now, define $\mathbf{Z}(t) = (Z_1(t), Z_2(t), \ldots, Z_{k-1}(t))$. The chi-square test for testing $H_0 : h_1 = h_2 = \ldots = h_k$ is then defined by

$$\chi^2(t) = \mathbf{Z}(t) \hat{\boldsymbol{\Sigma}}(t)^{-1} \mathbf{Z}(t), \tag{4.7}$$

where $\hat{\boldsymbol{\Sigma}}(t) = \{\hat{\sigma}_{lm}(t)\}$. Note that $\chi^2(t)$ follows chi-square distribution with $k - 1$ degrees of freedom.

The choice of weight function $K(t)$ can be differently specified for a rich class of test statistics. Recall that we truly have $K_l(t) = Y_l(t)K(t)$. As with the one-sample case, we obtain the logrank test for $K(t) = I(Y_\bullet > 0)$ whereby $K_l(t) = Y_l(t)$. Why is the test called as the logrank test? As explained in ABGK, Peto and Peto (1972)[90] argue that in the uncensored case, the logrank test becomes the savage test where the scores are approximately linearly related to the logarithm of the ranks of observations. The Gehan's test is obtained for $K(t) = Y_\bullet(t)$, and this times is also called as Gehan-Breslow test. The Tarone-Ware test is the result of choosing $K(t) = Y_\bullet^{1/2}(t)$. The Harrington-Fleming class of tests is the extension $K(t) = \left[\hat{S}(t-) \right]^\rho \left[1 - \hat{S}(t-) \right]^{1-\rho}$, for $0 \le \rho \le 1$. We will begin the discussion with the logrank test for the PBC dataset and then extend the other tests will be illustrated in the next example using the `logrank_test` function from the `coin` package.

Example 15 *The Log-Rank Test for the PBC Data. The logrank test is easily carried out for the PBC dataset by using the* `survdiff` *function. Recall from Examples 5 and 8 that we were not able to conclude if there was significant difference between the placebo and D-penicillamine. We now apply the logrank tests for inspecting whether the survival curves for the treatments are same, and whether there is a distinction by the gender. The following code chunk does the required.*

```
> # k-sample Nonparametric Tests
> data(pbc)
> # The Log-rank test
> survdiff(Surv(pbc$time,pbc$status==2)~trt,data=pbc)
Call:
survdiff(formula = Surv(pbc$time, pbc$status == 2) ~ trt,
        data = pbc)
n=312, 106 observations deleted due to missingness.
```

```
          N Observed Expected (O-E)^2/E (O-E)^2/V
trt=1 158       65      63.2     0.0502      0.102
trt=2 154       60      61.8     0.0513      0.102
```

```
 Chisq= 0.1  on 1 degrees of freedom, p= 0.7
> survdiff(Surv(pbc$time,pbc$status==2)~sex,data=pbc)
Call:
survdiff(formula = Surv(pbc$time, pbc$status == 2) ~ sex,
      data = pbc)
```

```
          N Observed Expected (O-E)^2/E (O-E)^2/V
sex=m  44       24      17.3     2.640       2.98
sex=f 374      137     143.7     0.317       2.98
```

```
 Chisq= 3  on 1 degrees of freedom, p= 0.08
```

The p-value for the treatment comparison is $p = 0.7$ *which means that there is not enough evidence in the data to reject the hypothesis that the treatments have equal effect. On the other hand, at* 10% *significance level, the male and female have significantly different response rates.* □

In the next example, we will apply the other forms of the k-sample tests to clinical trial data emerging in a kidney dialysis stury.

Example 16 *Various k-Sample Tests for Kidney Dialysis. Klein and Moeschberger (2003)[64] data emerging from a clinical trial which looks at the effectiveness of two methods of placing catheters in kidney dialysis patients. A catheter is the insertion of a flexible tube through a narrow opening in the body cavity. The first way of placing the catheter is a surgery, and the second technique is percutaneous treatment wherein a scope is inserted through an incision in the back to remove the kidney stones. The goal is to investigate if there is a difference in the time to cutaneous exit-site infection between patients whose catheter was placed surgically or by the percutaneous method. We have observation times, event indicator, and the type of treatment. The dataset is provided in the file* Kidney_Dialysis.csv *which is imported in the R session using the* read.csv *functions. We will first look at the survival curves under the two treatments.*

```
> # Performing Various k-Sample Tests
> KD <- read.csv("../Data/Kidney_Dialysis.csv",header=TRUE)
> head(KD)
  Time Indicator Treatment
1  1.5         1  Surgical
2  3.5         1  Surgical
3  4.5         1  Surgical
```

```
4   4.5          1  Surgical
5   5.5          1  Surgical
6   8.5          1  Surgical
> KD_KM <- survfit(Surv(Time,Indicator==1)~Treatment,data=KD)
> print(KD_KM)
Call: survfit(formula = Surv(Time, Indicator == 1) ~
               Treatment, data = KD)
                    n events median 0.95LCL 0.95UCL
Treatment=Percutaneous 76    11     NA      NA      NA
Treatment=Surgical     43    15   18.5    15.5      NA
> pdf("../Output/Kidney_Dialysis_Kaplan_Meier_Curves.pdf",
+     height=10,width=10)
> plot(survfit(Surv(Time,Indicator==1)~Treatment,data=KD),
+       conf.int = TRUE,col=c("red","blue"))
> legend(0,0.6,c("Percutaneous","Surgical"),pch="_",
+        col=c("red","blue"))
> dev.off()
RStudioGD
        1
```

*The Figure 4.1 shows the empirical survival curves for the two treatments
and it appears that the time to cutaneous exit-site infection for the percutaneous method is larger than for the surgical procedure. Let us apply the different survival tests and carry out the comparison. We use the* logrank_test
function from the coin *package.*

```
> # The Log-Rank Test
> logrank_test(Surv(Time,Indicator==1)~Treatmet,
+              type="logrank",data=KD)

Asymptotic Two-Sample Logrank Test

data:  Surv(Time, Indicator == 1) by
 Treatment (Percutaneous, Surgical)
Z = 1.6492, p-value = 0.0991
alternative hypothesis: true theta is not equal to 1

> # The Gehan Test
> logrank_test(Surv(Time,Indicator==1)~Treatment,
+              type="Gehan-Breslow",data=KD)

Asymptotic Two-Sample Gehan-Breslow Test

data:  Surv(Time, Indicator == 1) by
 Treatment (Percutaneous, Surgical)
Z = -0.046487, p-value = 0.9629
```

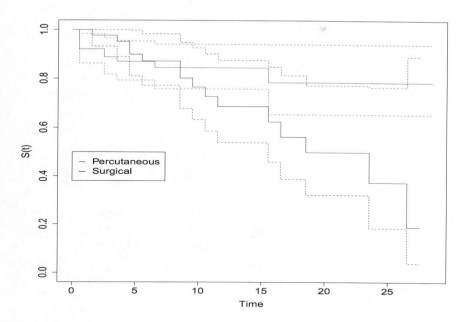

FIGURE 4.1: Kaplan-Meier Curves for Percutaneous and Surgical Kidney Dialysis Patients

```
alternative hypothesis: true theta is not equal to 1

> # The Tarone-Ware Test
> logrank_test(Surv(Time,Indicator==1)~Treatment,
+              type="Tarone-Ware",data=KD)

Asymptotic Two-Sample Tarone-Ware Test

data:  Surv(Time, Indicator == 1) by
 Treatment (Percutaneous, Surgical)
Z = 0.64996, p-value = 0.5157
alternative hypothesis: true theta is not equal to 1

> # The Fleming-Harrington with p=0,q=1
> pvalue(logrank_test(Surv(Time,Indicator==1)~Treatment,
+              type="Fleming-Harrington",rho=0,gamma=1,
+              data=KD))
[1] 0.0008613051
> # The Fleming-Harrington with p=1,q=0
> pvalue(logrank_test(Surv(Time,Indicator==1)~Treatment,
```

```
+                    type="Fleming-Harrington",rho=1,gamma=0,
+                    data=KD))
[1] 0.2249618
> # The Fleming-Harrington with p=1,q=1
> pvalue(logrank_test(Surv(Time,Indicator==1)~Treatment,
+                    type="Fleming-Harrington",rho=1,gamma=1,
+                    data=KD))
[1] 0.00085847
> # The Fleming-Harrington with p=0.5,q=0.5
> pvalue(logrank_test(Surv(Time,Indicator==1)~Treatment,
+                    type="Fleming-Harrington",rho=0.5,gamma=0.5,
+                    data=KD))
[1] 0.001299642
```

The logrank test, the Gehan test, and the Tarone-Ware test lead to the conclusion that the differences are not statistically significant. However, the p-values for many combinations of ρ in the Fleming-Harrington tests suggests that the treatment times are significantly different.

□

4.5   Exercises

Exercise 4.1 *Permutation Tests. For a two-sample test, the null hypothesis is the equality of survival functions $H_0 : S_1 = S_2$. A consequence of the assumption then is that the observation from one population is as likely to have come from the other population. Swapping an observation from one sample with an observation of the other sample, the test can be repeated. This exercise is repeated a large number of times and the conclusion is based on an aggregated value, say the average of the p-values. This is the well know permutation test. Create a dummy R function in executing the permutation tests for the logrank tests.*   □

Exercise 4.2 *Testing for Equality of Average Lifetimes of Two Treatments. In Section 3.2.3, the mean and median computations based on the Kaplan-Meier survival curves were addressed. We do not have a direct test for evaluation of equality of survival times. Now, carry out a nonparametric bootstrap test for equality of average lifetimes over two populations.*   □

Exercise 4.3 **Veteran's Lung Cancer Study.** *In the Veterans' Administration Lung Cancer study, available in the* **survival** *package, the purpose is to study the influence of the treatment among the veteran's response to the treatment expose to lung cancer. The goal is to find out if the new test treat-*

ment improves the lifetime compared with the standard treatment. Apply the two-sample logrank test for equality of survival curves.

4.6   Más Lejos Temas

In this chapter, we came across multiple comparison tests for comparison of equality of survival curves/functions. From parametric to nonparametric tests, logrank to weighted tests, the practitioner will find handful of techniques to apply them for analyzing survival data. The choice of the test procedure will remain a challenge and we usually have a consensus among the different techniques. Bootstrap extension of the techniques can be incorporated in a straightforward manner. Similarly, the delete-one-jackknife procedure is also a very effective method in carrying out the nonparametric tests. The delete-one-jackknife is a powerful method to determine the outliers too.

The methods in the chapter are useful under the critical assumption of noninformative censoring. Whenever the assumption is violated, or the censoring distribution plays a critical role too, we need to have techniques that extend approaches presented in this chapter. Robins and Rotnitzky (1992)[96] develop extensions of the logrank procedure in their papers Section 3. Quality of life is an important paradigm, as loosely mentioned in Section 2.7, and Health Related Quality of Life is a broader framework that also has scenarios where the censoring variable is informative. The comparison of survival curves of Quality Adjusted Lifetimes is dealt in Zhao and Tsiatis (2001)[121], which is weighted extension of the logrank test.

Statistical tests of hypotheses related to transition probability matrices of nonhomogeneous Markov process have been developed in Tattar and Vaman (2008)[114] and Tattar and Vaman (2014)[115]. The bootstrap version of the tests is availabe in Tattar (2016)[110].

Chapter 5

Regression Models

5.1 Introduction

In Chapters 3 and 4, we saw techniques to carry out estimation and testing of parameters, hazard rate, survival function, etc. Apart from the time to event and the event indicator, we made use of additional categorical variables such as the treatment and gender. While we are able to assess whether the survival curves across the treatments are same or not, we are not directly modeling the impact of the treatments on the survival curve or the lifetime. In the examples of the Mayo clinic PBC 1.2.2, bone marrow transplant data for leukemia 1.2.4, and the heart transplant monitoring data 1.2.5, we have a wealth of information in prognostic variables and demographic variables. Age, gender, height, weight, donor age, etc., are measurements that certainly influence the time to recover as well as the time to event of interest, while information on GVHD, platelet count and recovery, patient CMV status, the type of treatment, use of steroids, etc., make up information that directly intervene with the patient's health status and recovery abilities. Such data are not merely collected for records purpose, but to model the impact on clinical parameters. We saw complexities in simple parameters because of censoring and truncation. it's does not get simpler while modeling the impact of covariates.

Generally, regression analysis is the study undertaken to understand and model the influence of explanatory/exogenous/covariates on the variable of interest. In classical linear regression analysis, we directly model the influence of the covariates on the output/endogenous variable through linear coefficients. Typically, the model is $Y = \beta_0 + \beta_1 x_1 + \ldots + \beta_p x_p + \epsilon$, where $x_1, \ldots, x_p$ are the covariates, and $\beta_0, \beta_1, \ldots, \beta_p$ are regression coefficients. The notation of small case x need not be confused as the realized value of the state mapping function X introduced in Chapter 2. It is then tempting to model the lifetimes directly as a linear function of the covariates and the coefficients, say, $T = \beta_0 + \beta_1 x_1 + \ldots + \beta_p x_p + \xi \delta + \epsilon$, with T being the lifetime and δ the censoring indicator. The direct modeling of the covariates comes with a bag of limitations, and especially when the asymptotics get involved. However, we will begin modeling of the covariates with linear methods in spite of the limitations. The first technique is the Koul-Susarla-van Ryzin (1981)[66]

DOI: 10.1201/9781003306979-5

method where the least squares method is augmented with weights provided by the Kaplan-Meier estimates. Miller (1976)[81] proposed weighing the residuals with the Kaplan-Meier estimates and iterating until convergence. The authors are not aware of R packages which handle the estimation of the coefficients with these two methods and hence two new functions are created facilitating the computations. A third method is available in the Buckley-James estimator and we will resort to the R package rms. The problem of the lifetime predicted by the model being negative can be easily circumvented by using the log transformation. Remember, a subtle and understated assumption in the linear model $Y = \beta_0 + \beta_1 x_1 + \ldots + \beta_p x_p + \epsilon$ is that Y is a real variable that can take negative values as well. However, we have stability issues related with the coeffients when directly modeling the covariates linearly on the lifetimes.

Cox (1972)[28] provided the major breakthrough of covariate modeling for lifetime variable through an innovative relative risk modeling framework. Instead of modeling on the lifetimes, Sir D.R. Cox proposed to model the influence of the covariates on the hazard rate. The relative risk model, or the Cox proportional hazard model, is a semiparametric method in the sense that the baseline hazard rate is completely unspecified and it is any nonnegative function satisfying the conditions of the hazard rate. The covariates have a linear impact on the exponential scale and the finite dimensional parameters are parametric in nature, and hence the model is a semiparametric model. Justification for estimation of regressor parameters was provided by D R Cox in his seminal paper (1975)29) in the partial likelihood framework. We will delve in detail on this in Section 5.3. The residual analysis for the relative risk model will be performed in Section 5.4. We will close the chapter with a discussion of modeling the covariates for the lifetimes through parametric distributions in the Section 5.5.

5.2 Linear Regression Methods

The instability in the least-squares solution arises primarily because of the censored observations. A central idea in this section is then to 'complete' the censored observations in some sense, and proceed as with uncensored observations. As mentioned in the introduction, the two techniques in the direction of the linear regression methods for censored data are provided in Koul, et al. (1981)[66] and Miller (1976)[81]. We will consider one more method proposed in Buckley and James (1979)[21]. Comprehensive details over these techniques can be obtained in Chapters 10 and 11 of Smith (2002)[107].

We continue with the usual setup and denote the time to failure by T_i, the censoring variable C_i, observed time $\tilde{T}_i = T_i \wedge C_i$, $\delta_i = I\{\tilde{T}_i = T_i\}$, $i = 1, 2, \ldots, n$. Let $\mathbf{x}_i(t) = (x_{i1}(t), x_{i2}(t), \ldots, x_{ip}(t)), i = 1, \ldots, n$ be the covariate of the i-th observation at time t. However, in this section we will treat the covariates as time independent variables, that is, $\mathbf{x}_i = (x_{i1}, x_{i2}, \ldots, x_{ip})$, i.e., one or more of these covariates may vary with time. As before, we denote the distribution functions of T and C respectively by F and G. The preliminary assumption is that the observations $T_1, \ldots, T_n$ are mutually independent, and so are $C_1, \ldots, C_n$. Further, we assume that the T's and C's are conditionally independent given $\mathbf{x}$.

The observation $\tilde{T}_i$ can be written as follows:

$$\tilde{T}_i = \delta_i T_i + (1 - \delta_i) C_i, i = 1, \ldots, n. \tag{5.1}$$

The observation T_i may not be realized because of many reasons and we instead know the censored value C_i. We now plot the survival data points from Example 13 for the control group where the patients undergo only surgery. This treatment was extended to fifteen patients and seven out of them are censoring.

```
> # Plot of Survival Data
> CG_t <- c(0,2,5,7,12,42,46,54,7,11,19,22,30,35,39)
> CG_d <- rep(c(1,0),times=c(8,7))
> names(CG_t) <- paste0("Obsv.",1:15)
> pdf("../Output/Survival_Data_CG.pdf",height=15,width=15)
> dotchart(CG_t,xlim=c(0,60),cex = 1.5,pch=rep(c(4,1),
+           times=c(8,7)),main="Control Treatment")
> for(i in 1:length(CG_t)){
+   segments(0,i,CG_t[i],i,col="black")
+ }
> dev.off()
null device
          1
```

We do not have an outright function which will produce the plot of survival data as seen in Figure 1.2 of Moore (2016)[85], the plot on the cover design of Lee and Wang (2003)[71], etc. We will manipulate the regular output of a `dotchart` toward our end. The specification of the option of `pch` with the choice of 4 and 1 respectively lead to the symbols of × and ∘ on the dotchart. Now, the dotchart produces dashed line by default and we use `segments` option to produce a black line from the origin of time to the value of the observation `CG_t`.

Now, if all the datapoints in Figure 5.1 are marked as × indicating the observations of event of interest, we would have no problem in estimating the coefficients by using the least-squares method. The rate of convergence would be slower when the distribution function F is not normal, but we could still live with it. The complication arises because of the datapoints that end in ∘ in the visualization of the survival data for the top seven datapoints. In this section, we try to convert the ∘s into ×s and use the regular techniques to access the influence of the covariates on the lifetimes. Let's dive deeper in the Koul-Susarla-van Ryzin estimator.

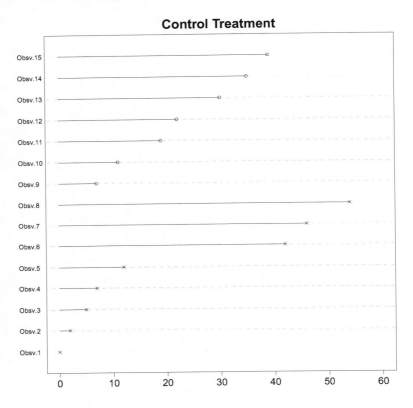

FIGURE 5.1: Visualizing Survival Data

To summarize, what is happening is the following. We want to understand the failure times T_i in terms of the associated prognostic and other related variables through a linear model:

$$T_i = \beta_0 + \beta_1 x_1 + \beta_2 x_2 + \ldots + \beta_p x_p + \epsilon, i = 1, \ldots, n.$$

Here, we have $T_i \sim F(= 1 - S)$, $\epsilon \sim F_\epsilon$ with $\sigma_\epsilon = \mathtt{Var}(\epsilon)$. Because of censoring variable, $C_i \sim G(= 1 - \bar{G})$, the observed values are $\tilde{T}_i = T_i \wedge C_i, \delta_i = I\{\tilde{T}_i = T_i\}$. Our interest is in carrying out inference regarding the coefficients $\boldsymbol{\beta}$ and the distribution function F.

The model for T_i specified above is sometimes referred to as *accelerated failure time (AFT) model*.

5.2.1 Koul-Susarla-van Ryzin Estimator

Koul, et al. (1981) considered the fundamental identity:

$$E\left[\frac{\delta_i \tilde{T}_i}{1 - G(\tilde{T}_i, \mathbf{x}_i)}|\mathbf{x}_i\right] = E\left[\frac{\delta_i \tilde{T}_i}{\bar{G}(\tilde{T}_i, \mathbf{x}_i)}|\mathbf{x}_i\right] = \beta'\mathbf{x}_i. \tag{5.2}$$

Here, $\bar{G} = 1 - G$ denotes the survival function of the censoring variables. Now, define $\check{T}_i = \delta_i \tilde{T}_i / \bar{G}(\tilde{T}_i, \mathbf{x}_i)$. Equation 5.2 shows that $\check{T}_i$ is linear in β. Now, if the probabilities $\bar{G}(\tilde{T}_i, \mathbf{x}_i)$ are known for the censored observations, the Koul-Susarla-van Ryzin technique replaces the censored observations by 0 and the event-realized observations $\tilde{t}_i$ by $\tilde{t}_i / \bar{G}(\tilde{t}_i, \mathbf{x}_i)$. Thus, 0 is the weight assigned to the censored observation and $\bar{G}(\tilde{T}_i, \mathbf{x}_i)$ for the event-realized observation. With these weights, the impact of the covariates on the lifetimes is obtained through the usual least-squares technique.

We are not aware of any R package or function which implements the technique. Hence, we create a function by the name KSvR which will carry out the necessary computations. Generally, δ values of 1 represents event being observed and 0 for censored. However, we need to find the survival probabilities $\bar{G}(\tilde{t}_i, .)$ based on $\tilde{t}_i, i = 1, \ldots, n$. The survival probabilities at the time points $\tilde{t}_i$ are obtained by using the `survfit` function with the censoring values as the events. Next, we obtain the weights at all observation times. The specification of the weights for the failure times and censoring times are taken care with `weights=delta*KSvR_weights`, and it is then passed to the linear regression model `lm`. The coefficients of the fitted linear model are then returned as the outpot of carrying out the Koul-Susarla-van Ryzin estimator. We first define the required function and then apply it for the PBC dataset.

```
> # Koul-Susarla-van Ryzin Estimator
> KSvR <- function(times,delta,covnames,data){
+    # Survival function for the censoring variables
+    delta2 <- (delta==0)
+    GC <- survfit(Surv(times,delta2)~1)
+    # Weights for the Koul-Susarla-van Ryzin estimator
+    KSvR_weights <- summary(GC,times)$surv
+    lm1 <- lm(as.formula(
+      paste("times~",paste0(covnames,collapse="+"))),
+      weights=delta*KSvR_weights,data)
+    return(coef(lm1))
+ }
```

Example 17 *Fitting the Koul-Susarla-van Ryzin Estimator for the Primary Biliary Cirrhosis Data. We are now ready to use the* KSvR *function on the PBC dataset. Note that we do not have a formula type of specification in the function. It is generated later in the module by using* as.formula, *and the content of the text matter is combined with the variable names using the* paste *function. The KSvR is now ready for action. We* *will first do a preliminary visualization of the survival time against the four features/covariates—age, edema, bilirubin, and albumin.*

```
> # Applying to PBC
> data(pbc)
> covnames <- c("age","edema","bili","albumin")
> pdf("PBC_Matrix_Scatter_Plots.pdf",height=20,width=20)
> plot(pbc[,c(2,5,10,11,13)],main="PBC Matrix of Scatter Plots")
> dev.off()
RStudioGD
        1
> KSvR(pbc$time,pbc$status==2,covnames = covnames,data=pbc)
```

The `plot` *function gives the matrix of scatterplots. For simplicity, we ignore the failure time indicators. Now, Figure 5.2 shows, a bit subjectively speaking, that on average as age increases, time to event decreases. Similar statement holds for edema and bilirubin. However, time to death increases in albumin. Does the Koul-Susarla-van Ryzin estimate indicate the same?*

Interpretation of the coefficients is straightforward as in the general linear regression. Age is expectedly having a negative impact on the survival time. The coefficients of the other variables validate the findings from the matrix of scatterplot. □

A word of caution. Looking at the use of the `lm` function, or the typical Gaussian linear model, it might be tempting to extract other related parameter values such as R^2, p-values, confidence intervals, etc. However, it is not advisable to do so. The sole purpose of the `lm` function was to use the least-squares principle of obtaining βs. How does one then obtain the confidence interval and other parameters? We will not delve into more details for the Koul-Susarla-van Ryzin estimator, or other methods in this section. The only strategy to obtain the solution then is the recourse to the resampling methods, and especially the bootstrap method.

5.2.2   Miller's Estimator

Miller (1976)[81] proposed estimation of the regression coefficients in the linear model by using Kaplan-Meier estimator for the residual distribution function. Consider the linear model again $T_i = \beta_0 + \beta_1 x_1 + \ldots + \beta_p x_p + \epsilon$. We have $\epsilon \sim F_\epsilon$. Miller suggests use of weighted least squares to estimate β by using the Kaplan-Meier estimates of the residuals of the linear model. That is, in the first iteration β is obtained using ordinary least squares. The residuals are calculated by $r_i = \hat{t}_i - \tilde{t}_i, i = 1, \ldots, n$. Now, using $r_1, \ldots, r_n$ and $\delta_1, \ldots, \delta_n$, the Kaplan-Meier estimator for the distribution, $F_\epsilon = 1 - S_\epsilon$ is obtained. Note that some residuals, r_i's, are likely to be negative. Thus, even as we always

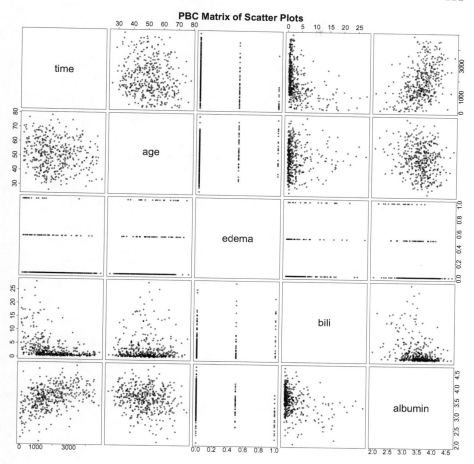

FIGURE 5.2: Matrix of Scatterplots

fit the Kaplan-Meier estimator for nonnegative $\tilde{t}$s, in the case of the residuals, we simply use the technique for obtaining the empirical distribution function of F_ϵ. Now, using the fitted Kaplan-Meier estimator $\hat{S}_\epsilon = 1 - \hat{F}_\epsilon$, we obtain the survival probabilities for $\hat{S}_\epsilon(r_1), \ldots, \hat{S}_\epsilon(r_n)$ and pass them as weights to the model $T_i = \beta_0 + \beta_1 x_1 + \ldots + \beta_p x_p + \epsilon$ to obtain $\boldsymbol{\beta}$. We keep on iterating the steps over and again until the norm between the successive iterations of the regression coefficients become almost zero.

As with the Koul-Susarla-van Ryzin estimator, we are not aware of an R implementation of the Miller's estimator. The `Miller` function carries out the steps described previously. The functions `lm`, `coefficients`, `residuals`, `survfit` and `paste` are useful in setting up the `Miller` function.

```
> # The Miller's Weighted OLS Regression
> Miller <- function(times,delta,covnames,data){
+    lm1 <- lm(as.formula(
+      paste("times~",paste0(covnames,collapse="+"))),data)
+    norm_incre <- 5
+    while(norm_incre>1e-4){
+      old_coef <- coefficients(lm1)
+      rt <- residuals(lm1) # the residual times
+      tempfit <- survfit(Surv(rt,delta)~1)
+      KMW <- summary(tempfit,times = rt)$surv
+      # Kaplan-Meier Weights
+      lm1 <- lm(as.formula(
+        paste("times~",paste0(covnames,collapse="+"))),
+        weights=KMW,data)
+      new_coef <- coefficients(lm1)
+      norm_incre <- sqrt(sum((old_coef-new_coef)^2))
+    }
+    return(new_coef)
+ }
```

We next apply it to the PBC dataset with the same set of variables as in Example 17.

Example 18 *Fitting the Miller's Estimator for the Primary Biliary Cirrhosis Data. Contd. The* `Miller` *function is now used to obtain the coefficients of the variables.*

```
> Miller(pbc$time,pbc$status==2,covnames = covnames,data=pbc)
(Intercept)         age        edema       bili
 -1,8.765035  -7.293841  -478.503310  -76.346909
     albumin
 772.236754
```

The signs of the coefficients in this technique are aligned with the Koul-Susarla-van Ryzin coefficients as seen in Example 17.   □

The statistical inference of the coefficients can be carried out using the resampling methods. We will next move to a technique which is probably more popular in directly modeling the influence of the covariates on the lifetimes.

5.2.3 Buckley-James Estimator

Weights were used to improve the estimators in the two previous subsections. The censored observations receive zero weights in the Koul-Susarla-van Ryzin estimator while the event observation times are assigned weights with the probability of not being censored $\bar{G}(.)$ at the time of the event being observed. In the Miller's technique too, the weights are determined, iteratively till convergence. The Buckley-James estimator is another alternative for direct modeling of the covariates on the failure times.

Recollect the relationship between T_i, C_i and $\tilde{T}_i$ as $\tilde{T}_i = T_i\delta_i + C_i(1 - \delta_i)$. Buckley and James (1979)[21] suggested to replace the censored observations C_i, the cases when $\delta_i = 0$, by $E(T_i|T_i > t)$. A new variable is now defined by

$$V_i = T_i\delta_i + E(T_i|T_i > t)(1 - \delta_i), i = 1, \ldots, n. \tag{5.3}$$

It can be then proved that $E(V_i) = \boldsymbol{\beta}'\mathbf{x}_i$. In this method, the censored values are replaced by the corresponding conditional expected values.

The procedure for obtaining the regression coefficients $\boldsymbol{\beta}$ is as follows. First, begin by an initial solution $\mathbf{b}$ of the regression coefficients. Calculate the residuals $e_i(\mathbf{b}) = \tilde{t}_i - \mathbf{b}'\mathbf{x}_i$. We obtain the Kaplan-Meier estimator based on the residuals e_i and denote it by $\hat{S}(e_j)$. The **Buckley-James estimator** of $\boldsymbol{\beta}$ is the solution of the following equations:

$$\sum_{i=1}^{n} \left\{ \delta_i e_i(\mathbf{b}) + (1 - \delta_i) \sum_{j:e_j>e_i} \frac{e_i(\mathbf{b})\Delta\hat{S}(e_j)}{1 - \hat{S}(e_i)} \right\} \mathbf{x}_i = 0. \tag{5.4}$$

The summation in this equation $\sum_{j:e_j>e_i} \frac{e_i(\mathbf{b})\Delta\hat{S}(e_j)}{1-\hat{S}(e_i)}$ is an estimate of the conditional expectation $\hat{E}(T_i - \mathbf{b}'\mathbf{x}_i|\tilde{t}_i, \boldsymbol{\beta} = \mathbf{b}, \delta_i = 0)$. Luckily, we do not have to worry over writing complex codes to obtain the Buckley-James estimator. The rms package contains a function bj, standing for Buckley-James, which does the required computations. We apply the function on the PBC data next.

Example 19 *Fitting the Buckley-James Estimator for the Primary Biliary Cirrhosis Data. Contd. The* bj *function can be used in a straight-forward manner.*

```
> # The Buckley-James Estimator
> BJ_pbc <- bj(Surv(pbc$time,pbc$status==2)~age+edema+bili+
+            albumin,link="identity",data=pbc)
> coef(BJ_pbc)
Intercept       age      edema       bili    albumin
 2139.08    -36.14    -855.68    -140.46     913.33
> Miller(pbc$time,pbc$status==2,covnames = covnames,data=pbc)
(Intercept)           age      edema        bili    albumin
   -18.765        -7.294   -478.503     -76.347    772.237
> KSvR(pbc$time,pbc$status==2,covnames = covnames,data=pbc)
(Intercept)           age      edema        bili    albumin
    799.96        -12.97   -574.91      -52.74     532.70
```

The Buckley-James coefficients provide similar insights as the other two methods. The theory of the Buckley-James estimator is well developed and we can obtain the p-values and confidence intervals using the inbuilt options with the bj *function. We next obtain these quantities in R.*

```
> BJ_pbc
Buckley-James Censored Data Regression
 bj(formula = Surv(pbc$time, pbc$status == 2) ~ age + edema +
    bili + albumin, data = pbc, link = "identity")
                                     Discrimination
                                        Indexes
 Obs     418     Regression d.f.4     g      1097.017
 Events 161      sigma1137.2283
                 d.f.          156

               Coef      S.E.      Wald Z Pr(>|Z|)
 Intercept 2139.0824 975.7036    2.19   0.0284
 age         -36.1431   9.6216  -3.76   0.0002
 edema      -855.6754 290.4991  -2.95   0.0032
 bili       -140.4601  16.7816  -8.37   <0.0001
 albumin     913.3305 213.4455   4.28   <0.0001

> confint(BJ_pbc)
                2.5 %   97.5 %
 Intercept    226.7   4051.43
 age          -55.0    -17.29
 edema      -1425.0   -286.31
 bili        -173.4   -107.57
 albumin      495.0   1331.68
```

The p-values for all the coefficients is less than 0.05 *and hence we can conclude that they are significant in explaining the overall lifetime in the PBC
study.*   □

When interpretable results are required, the Buckley-James estimator offers a rare approach in survival analysis in which we can directly call out the
influence of the covariates on the failure times. With strong theoretical back-
ing going for it, the Buckley-James estimator will often come handy. However,
regression analysis is predominantly carried out in the area using the Cox relative risk model. It is sometimes popularly known as the Cox proportional
hazards model. The proportional hazards model is a particular case of the
relative risk model and we will take up the discussion and development in the
next section.

5.3   Relative Risk (Cox) Model

The Kaplan-Meier estimator and the Nelson-Aalen estimator are used more
widely than the parametric models. While the parametric approach has seen

further extensions in mixed parametric distributions and generalized lifetime distributions, nonparametric methods continue to be applied more frequently and the gap between them has not shrunk. As we had seen in the previous section, we have three methods of direct modeling of the covariate impact on the failure times, and the Buckley-James estimator is more widely used among them. However, in real practice, the techniques are rarely used. Much before that, the attempts to generalize the usual least squares had not seen success. The problems persisted mainly with absence of desirable statistical properties and most results were unstable. Cox (1972)[28] achieved the break-through in the landmark paper *Regression Models and Life Tables*. This paper has over fifty-five thousand citations. Here, the influence of the covariates is modeled through a risk function on the hazard rate and the baseline hazard. The covariates have a finite p dimensional specification, while the baseline is left completely unspecified and the space of functions is infinite-dimensional in nature. Hence, the nomenclature of semiparametric model is apt for the Cox relative risk model. We will now get in the details of the model.

Consider n patients in a study and let $N_i(t), i = 1, \ldots, n$ be the counting process which counts the event occurrences in the time interval $[0, t]$. The counting process N_i is associated with the lifetime $T_i, T_i \sim F = 1 - S$, and censoring variable $C_i, C_i \sim G$, and we have $\tilde{T}_i = T_i \wedge C_i, \delta_i = I(\tilde{T}_i = T_i, \forall i = 1, \ldots, n$. Let $h(t)$ be the hazard rate associated with S. With $Y_i(t)$ denoting the at-risk process associated with the i-th observation. We denote the i-th covariate vector by $\mathbf{x}_i(t) = (x_{i1}(t), x_{i2}(t), \ldots, x_{ip}(t))$. With covariates, the hazard rate of the i-th observation is expressed as $h_i(t|\mathbf{x}_i(t))$. The intensity process $h_i(t)$ associated with the counting process $N_i, i = 1, \ldots, n$ can be expressed as

$$h_i(t) = Y_i(t)h_i(t|\mathbf{x}_i(t)). \tag{5.5}$$

The intensity process $h_i(t)$ is at-risk process times the hazard rate which ensures that the effective hazard rate is essentially zero whenever the observation is not at risk of event occurrence. Strictly speaking, we should have another symbol for the intensity process, which activates the hazard rate according to the at-risk indicator variable. However, it would be often clear from the context whether we are talking about hazard rate or the intensity process.

The specification of the hazard rate $h_i(t|\mathbf{x}_i(t))$ determines the type of model we are engaged with. We consider the relative risk model here. Let $r(\boldsymbol{\beta}, \mathbf{x}_i(t))$ denote the relative risk function for the i-th observation, with $\boldsymbol{\beta} = (\beta_1, \beta_2, \ldots, \beta_p)$ being the vector of regression coefficients.

The reader might argue that if the covariates $\mathbf{x}_i(t) = (x_{i1}(t), x_{i2}(t), \ldots, x_{ip}(t))$ are time-dependent, the specification of $(\beta_1, \beta_2, \ldots, \beta_p)$ might be restrictive. In many settings and trials, even though the covariates might vary in time, their influence is seen to depend on their magnitude and not so much on time. However, in other scenarios, the time-independent coefficients are a compromised solution and it is required for us to specify the regression coefficients to be functions of time. We will consider such a model in a different framework though, see Section 6.4.

For an individual with covariate vector $\mathbf{x}_i(t)$, we consider the following form of relationship between the covariate and the hazard rate:

$$h_i(t|\mathbf{x}_i) = h_0(t)r(\boldsymbol{\beta}, \mathbf{x}_i(t)). \tag{5.6}$$

Here, the baseline $h_0(t)$ is left unspecified and the relative risk function $r(\boldsymbol{\beta}, \mathbf{x}_i(t))$ needs further specification. Model 5.6 is called as the *relative risk model*. The choice of $r(\boldsymbol{\beta}, \mathbf{x}_i(t)) = \exp\{\boldsymbol{\beta}'\mathbf{x}_i(t)\}$ leads to *Cox proportional hazard (PH) model* and it is given by

$$h_i(t|\mathbf{x}_i(t)) = h_0(t)\exp\{\boldsymbol{\beta}'\mathbf{x}_i(t)\}. \tag{5.7}$$

Cox PH model is also called as exponential relative risk model. Why proportional hazards? Consider two observations $\mathbf{x}_1(t)$ and $\mathbf{x}_2(t)$ where all the components are same except the l-th component and we have $x_{2l}(t) = x_{1l}(t) + 1$. Then the hazards ratio based on the model 5.7 becomes

$$\frac{h_2(t|\mathbf{x}_2(t))}{h_1(t|\mathbf{x}_1(t))} = \exp\left\{\boldsymbol{\beta}'\left(\mathbf{x}_2(t) - \mathbf{x}_1(t)\right)\right\} = \exp\left\{\beta_l\right\}.$$

Clearly, the hazards ratio is now independent of time and given the hazard rate of observation 1 at time t by $h_1(t|\mathbf{x}_1(t))$, we can obtain the second observation $h_2(t|\mathbf{x}_2(t))$ by multiplying the first by a scale of e^{β_l}. Thus, the model is aptly called as the *proportion hazards model*. The intensity process associated with the Cox PH model is thus given by

$$h_i(t) = Y_i(t)h_0(t)\exp\{\boldsymbol{\beta}'\mathbf{x}_i(t)\}, \tag{5.8}$$

and for the general relative risk model by

$$h_i(t) = Y_i(t)h_0(t)r(\boldsymbol{\beta}, \mathbf{x}_i(t)). \tag{5.9}$$

We briefly discuss other forms of the relative risk function. The relative risk function can be specified differently too. The two choices mentioned in ABG are the *linear relative risk function* $r(\boldsymbol{\beta}, \mathbf{x}_i(t)) = 1 + \boldsymbol{\beta}'\mathbf{x}_i(t)$, and the *excess relative risk function* $r(\boldsymbol{\beta}, \mathbf{x}_i(t)) = \prod_{l=1}^{p}\{1 + \beta_l x_{il}(t)\}$.

For the time-independent covariates, $\mathbf{x} = \mathbf{x}(t), \forall t$, the survival function S can be expressed as

$$S(t|\mathbf{x}) = S_0(t)^{\exp\{\boldsymbol{\beta}'\mathbf{x}\}},$$

where $S_0(t) = \exp\{-\int_0^t h(s)ds\}$. The class of models $[\{S_0(t)\}^c, 0 < c < \infty]$, where c is a constant, is referred as the *Lehmann family*.

Finding an estimate of the regression coefficients $\boldsymbol{\beta}$ requires us to find a way of handling the unknown baseline hazard rate $h_0(t)$. However, we do not have any information on the form or distribution of the hazard rate. Cox provided a workaround for the baseline hazard rate, and later justified the same in another benchmark paper *Partial Likelihood* in 1975.

For the n observations, the aggregated counting process is $N(t) = N_{\bullet}(t) = \sum_{i=1}^{n} N_i(t)$, and the associated intensity process is $h(t) = h_{\bullet}(t) = \sum_{i=1}^{n} h_i(t) = \sum_{i=1}^{n} r(\boldsymbol{\beta}, \mathbf{x}_i(t))$. ABG introduce factorization based on the intensity process which helps setting up the likelihood function. The intensity process of an observation i can be written as

$$h_i(t) = h(t)\pi(i|t),$$

where

$$\pi(i|t) = \frac{h_i(t)}{h(t)} = \frac{Y_i(t)r(\boldsymbol{\beta}, \mathbf{x}_i(t))}{\sum_{i=1}^{n} Y_i(t)r(\boldsymbol{\beta}, \mathbf{x}_i(t))}. \qquad (5.10)$$

Note that the baseline hazard rate gets canceled out in the above expression. The factor $\pi(i|t)$ has the interpretation of being the conditional probability of observing the event for individual i at time t given the history and the information that the event is observed at the time.

The partial likelihood function for $\boldsymbol{\beta}$ is obtained by multiplying the conditional probabilities (5.10) over all the observed failure times, that is $\delta = 1$ observations. As earlier, let $T_1 < T_2 < \ldots$ denote the times for which $\delta = 1$. Let m_i be the index of the individual who experiences an event at time T_i. Then, the *partial likelihood* is given by the following:

$$L(\boldsymbol{\beta}|\texttt{data}) = \prod_{T_i} \frac{Y_{m_i}(T_i)r(\boldsymbol{\beta}, \mathbf{x}_{m_i}(T_i))}{\sum_{l=1}^{n} Y_l(T_i)r(\boldsymbol{\beta}, \mathbf{x}_l(T_i))}.$$

Now, define the risk set at time T_i by $\mathcal{R}_i = \{o|Y_o(T_i) = 1\}$. The partial likelihood function can then be written in the popular form as follows:

$$L(\boldsymbol{\beta}|\texttt{data}) = \prod_{T_i} \frac{r(\boldsymbol{\beta}, \mathbf{x}_{m_i}(T_i))}{\sum_{l \in \mathcal{R}_i} r(\boldsymbol{\beta}, \mathbf{x}_l(T_i))}. \qquad (5.11)$$

The *maximum partial likelihood estimator* $\hat{\boldsymbol{\beta}}$ is the value of $\boldsymbol{\beta}$ that maximizes the partial likelihood (5.11). The maximum partial likelihood estimator $\hat{\boldsymbol{\beta}}$ enjoys the large sample properties of the general MLE, especially, $\hat{\boldsymbol{\beta}} \sim AN(0, \mathcal{I}(\hat{\boldsymbol{\beta}})^{-1})$, where $\mathcal{I}(\hat{\boldsymbol{\beta}})$ is the *information matrix*. The reader can refer Chapter 4 of ABG or Chapter 4 of Kalbfleisch and Prentice (2002)[60] for proof of this result.

Let $U(\boldsymbol{\beta}) = \frac{\delta}{\delta\boldsymbol{\beta}} \log L(\boldsymbol{\beta})$ denote the *score function*. The MLE is a solution of the equation obtained by equating the score function to 0. To test the null hypothesis $H_0 : \boldsymbol{\beta} = \boldsymbol{\beta}_0$, we have three options:

- *likelihood ratio test statistic*: $\chi^2_{LR} = 2\{\log L(\hat{\boldsymbol{\beta}}) - \log L(\boldsymbol{\beta}_0)\}$.

- *Rao's score test statistic*: $\chi^2_{SC} = U(\boldsymbol{\beta}_0)'\mathcal{I}(\boldsymbol{\beta}_0)^{-1}U(\boldsymbol{\beta}_0)$.

- *Wald test statistic:* $\chi_W^2 = (\hat{\boldsymbol{\beta}} - \boldsymbol{\beta}_0)' \mathcal{I}(\boldsymbol{\beta}_0)(\hat{\boldsymbol{\beta}} - \boldsymbol{\beta}_0).$

All the three statistics χ_{LR}^2, χ_{SC}^2, and χ_{SC}^2 are approximately distributed as χ^2 distribution with p degrees of freedom for sufficiently large n. Further, the three techniques are asymptotically equivalent. The reader may refer ABG and ABGK for further details related to the asymptotics.

We now return to the problem of estimating the baseline hazard rate. As in the earlier case, we will rather focus on estimating the baseline cumulative hazard function $H_0(t) = \int_0^t h_0(s)ds$. Recollect that the aggregate intensity process is of the form:

$$h(t) = h_\bullet(t) = \left(\sum_{i=1}^n Y_i(t) r(\boldsymbol{\beta}, \mathbf{x}_i(t)) \right) h_0(t).$$

It is clear from the expression that known value of regression coefficient vector $\boldsymbol{\beta}$, the baseline cumulative hazard function can be estimated by using the aggregated counting process $N(t) = N_\bullet(t)$ as follows:

$$\hat{H}_0(t|\boldsymbol{\beta}) = \int_0^t \frac{dN(s)}{\sum_{i=1}^n Y_i(t) r(\boldsymbol{\beta}, \mathbf{x}_i(s))}.$$

Substituting the unknown vector of regression coefficient $\boldsymbol{\beta}$ by the MLE, we get:

$$\hat{H}_0(t) = \int_0^t \frac{dN(s)}{\sum_{i=1}^n Y_i(t) r(\hat{\boldsymbol{\beta}}, \mathbf{x}_i(s))} = \sum_{T_i \leq t} \frac{1}{\sum_{m \in \mathcal{R}_i} r(\hat{\boldsymbol{\beta}}, \mathbf{x}_m(T_i))}. \tag{5.12}$$

The estimator 5.12 is the famous *Breslow estimator*.

Given a new observation with time-independent covariate $\mathbf{x}_0$, how does one obtain the required cumulative hazard function $\hat{H}_0(t|\mathbf{x}_0)$ and the survival function $\hat{S}_0(t|\mathbf{x}_0)$? It can be easily seen from the hazard rate specification of $h(t|\mathbf{x}, \boldsymbol{\beta}) = r(\boldsymbol{\beta}, \mathbf{x})h_0(t)$ that

$$H(t|\mathbf{x}_0) = \int_0^t r(\boldsymbol{\beta}, \mathbf{x}_0(s))h_0(s)ds.$$

Clearly, the cumulative hazard function $H(t|\mathbf{x}_0)$ can be then estimated by

$$\hat{H}(t|\mathbf{x}_0) = \int_0^t r(\hat{\boldsymbol{\beta}}, \mathbf{x}_0(s))d\hat{H}_0(s) = \sum_{T_i \leq t} \frac{r(\hat{\boldsymbol{\beta}}, \mathbf{x}_0(T_i))}{\sum_{m \in \mathcal{R}_i} r(\hat{\boldsymbol{\beta}}, \mathbf{x}_0(T_m))}. \tag{5.13}$$

Similarly, the survival function for the unknown observation $\mathbf{x}_0$ is given by

$$\hat{S}(t|\mathbf{x}_0) = \prod_{T_i \leq t} \left\{ 1 - \Delta \hat{H}(T_i|\mathbf{x}_0) \right\}. \tag{5.14}$$

We will now take the R action on the PBC dataset for these developments. First, we will compare the logrank test with the Cox PH, model.

Example 20 *Comparing the PBC Treatments Under the Cox Proportional Hazards Model. In Example 15, we had run the logrank test, for comparison of survival cuves of two samples, for identifying the response to the PBC recovery was equal across the treatments and the gender. The differences were not significant for the treatment while females were observed to have significantly higher lifetimes. We will repeat the output of the logrank test for the treatments and then using the formula* Surv(time,status==2)~trt *along with the* coxph *function, model the treatment effect on the time to failure in the PBC dataset.*

```
> # The Log-rank test and Cox PH Model
> data(pbc)
> # The Log-rank test
> survdiff(Surv(pbc$time,pbc$status==2)~trt,data=pbc)
Call:
survdiff(formula = Surv(pbc$time, pbc$status == 2) ~ trt,
        data = pbc)
n=312, 106 observations deleted due to missingness.

          N Observed Expected (O-E)^2/E (O-E)^2/V
trt=1 158       65     63.2    0.0502     0.102
trt=2 154       60     61.8    0.0513     0.102

 Chisq= 0.1  on 1 degrees of freedom, p= 0.7
> pbc_trt_ph <- coxph(Surv(time,status==2)~trt,data=pbc)
> summary(pbc_trt_ph)
Call:
coxph(formula = Surv(time, status == 2) ~ trt, data = pbc)

  n= 312, number of events= 125
   (106 observations deleted due to missingness)

        coef exp(coef) se(coef)      z Pr(>|z|)
trt -0.05722   0.94438  0.17916 -0.319    0.749

    exp(coef) exp(-coef) lower .95 upper .95
trt    0.9444      1.059    0.6647     1.342

Concordance= 0.499,  (se = 0.025 )
Likelihood ratio test= 0.1  on 1 df,    p=0.7
Wald test         = 0.1  on 1 df,    p=0.7
Score (logrank) test = 0.1  on 1 df,    p=0.7
```

The p-value associated with the treatment is 0.749 *which means that the D-penicillamine does not offer advantage over the placebo treatment. Similarly, the p-values associated with the chi-square tests χ^2_{LR}, χ^2_{SC}, and χ^2_W all are* *equal to* 0.7 *and lead to the equal to conclusion regarding the effectiveness of the treatment.*

$\square$

In the previous example, we simply used the Cox proportional hazards model for modeling the treatment effect. We will continue applying the Cox proportional hazards model in presence of more than one covariate.

Example 21 *Analysis of PBC Data Under the Cox Proportional Hazards Model. We are now interested to see how the variables of age, treatment, gender, edema, bilirubin, and albumin influence the hazard rate for the patients. We first subset the dataset for these variables and then use it to model the relative risk model. Applying the* summary *function, we obtain the p-values of the related covariates and we reproduce in the following code block.*

Note that in the pbc *dataset, we have 161 patients for which the event is observed,* status==2. *However, we have missing data in some of the covariates of interest and we need to remove them first to carry out the analysis.*

```
> # Modeling PBC with Cox PH Model
> pbc2 <- subset(pbc,select=c(time,status,trt,age,sex,
+                             edema,bili,albumin))
> pbc2 <- na.omit(pbc2)
> pbc_ph <- coxph(Surv(time,status==2)~trt+age+sex+
+                 edema+bili+albumin,data=pbc2)
> summary(pbc_ph)
Call:
coxph(formula = Surv(time, status == 2) ~ trt + age + sex +
    edema + bili + albumin, data = pbc2)

  n= 312, number of events= 125

              coef exp(coef)  se(coef)      z Pr(>|z|)
trt       -0.01428   0.98583   0.18565  -0.08  0.93871
age        0.03162   1.03212   0.00941   3.36  0.00077 ***
sexf      -0.57819   0.56091   0.24772  -2.33  0.01960 *
edema      1.02215   2.77917   0.31696   3.22  0.00126 **
bili       0.12549   1.13371   0.01485   8.45  < 2e-16 ***
albumin   -1.22733   0.29307   0.23997  -5.11  3.1e-07 ***
---
Signif. codes:  0 '***' 0.001 '**' 0.01 '*' 0.05 '.' 0.1 ' ' 1

          exp(coef) exp(-coef) lower .95 upper .95
trt           0.986      1.014     0.685     1.418
age           1.032      0.969     1.013     1.051
sexf          0.561      1.783     0.345     0.912
edema         2.779      0.360     1.493     5.173
```

```
bili        1.134      0.882      1.101      1.167
albumin     0.293      3.412      0.183      0.469


Concordance= 0.823   (se = 0.02 )
Likelihood ratio test= 158   on 6 df,    p=<2e-16
Wald test            = 183   on 6 df,    p=<2e-16
Score (logrank) test = 280   on 6 df,    p=<2e-16
```

```
> save(pbc_ph,file="../Output/pbc_ph.RData")
```

The estimates of the regression coefficients corresponding to covariates treatment, age, gender, edema, bilirubin, and albumin are respectively -0.01428, 0.03162, -0.57819, 1.02215, 0.12549, and -1.22733. The estimates of the standard errors are also provided in the column se(coef).

We can see that except trt *variable, all others are significant here because of lower p-value. The p-values of the various chi-square tests imply that we have to reject the null-hypothesis that all the variables have insignificant impact on the hazard rate.*

Let us now consider the baseline hazard rate $h_0(t)$ specified in Equation 5.7, and we know that its estimate is given by the Breslow estimator Equation 5.12. Using the R function basehaz *on the fitted Cox proportional hazards object, we easily obtain the Breslow estimator.*

```
> # Baseline Hazard
> pbc_baseline <- basehaz(pbc_ph)
> head(pbc_baseline)
    hazard time
1 0.001016   41
2 0.002199   51
3 0.003436   71
4 0.004685   77
5 0.005967  110
6 0.007252  130
> pdf("../Output/PBC_Baseline_Hazards.pdf")
> plot(pbc_baseline$time,pbc_baseline$hazard,"l",
+      xlab = "Time",ylab="h(t)",main="Baseline Hazard")
> dev.off()
RStudioGD
        2
```

The plot of the Breslow estimator is given in Figure 5.3.

☐

Now, for given two patients, we would like to plot their survival curves as given by expression 5.14.

Baseline Hazards

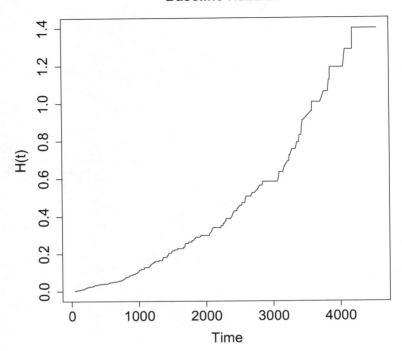

FIGURE 5.3: Breslow Estimator for the PBC Data

Example 22 *Survival Function For PBC*. *To fit survival curves for two patients with varied history, we will create two new data objects with two different set of covariate values. Next, we will combine them in a new matrix using the* rbind *function. This object is next passed to the* survfit *function along with the fitted Cox proportional hazards model.*

```
> new_patient_01 <- data.frame(trt=1,age=50,sex="f",
+                              edema=1,bili=5,albumin=3)
> new_patient_02 <- data.frame(trt=1,age=50,sex="m",
+                              edema=1,bili=5,albumin=3)
> nps <- rbind(new_patient_01,new_patient_02)
> pdf("../Output/Survival_Function_with_Cox_PH.pdf")
> plot(survfit(pbc_ph,newdata = nps),col=c("red","blue"),
+      ylab = "Survival Function",xlab= "Time",
+      main="Cox PH Fit for Two Patients")
> dev.off()
RStudioGD
    2
```

The Kaplan-Meier curve in red belongs to a female while the blue colored curve corresponds to a male, see Figure 5.4. We see that females have more expected lifetime than males.

□

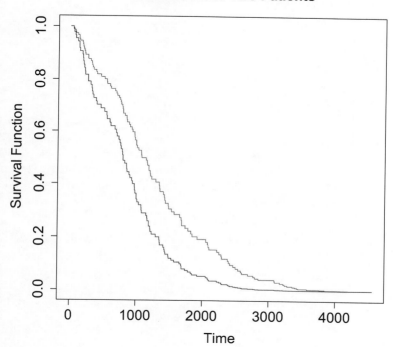

FIGURE 5.4: Survival Plots with Cox Model

5.4 Residual Analysis for the Cox Proportional Hazards Regression Model

Residual analysis plays a paramount role in diagnostics in regression analysis. In the classical regression analysis, with a slight and local abuse of notation for not confusing the at-risk process with Y's, we obtain the fitted values $\hat{Y}_1, \hat{Y}_2, \ldots, \hat{Y}_n$ corresponding to the actual values $Y_1, Y_2, \ldots, Y_n$, we have the residuals $r_i = Y_i - \hat{Y}_i$, $i = 1, 2, \ldots, n$. The residuals therein are useful for further analysis and we have influential points and leverage points, DFFITS, DFBETAS, and the Cooks distance as important measures, see Cook and Weisburg (1982)[27]. However, we do not have $\hat{T}_i$ as an output of the fitted Cox relative risk models. We obtain estimates of the regression coefficients and fit the baseline hazards function. Let us see how these quantities help us in obtaining the following type of residuals:

- Martingale Residuals

- Deviance Residuals

- DFBETAS

- Score Residuals

- Schoenfeld Residuals

Therneau and Grambsch (2000)[116] discuss variety of residuals in Chapter 4 of their book. We will begin with the idea of *martingale residuals*. The intensity process of the i-th observation under the relative risk model is $h_i(t) = Y_i(t)h_0(t)e^{\beta' x_i(t)}$, and substituting it in the martingale equation $M_i(t) = N_i(t) - H_i(t)$, we get

$$M_i(t) = N_i(t) - H(t) = N_i(t) - \int_0^t Y_i(s)h_0(s)e^{\beta' x_i(s)}ds, i = 1, 2, \ldots, n. \quad (5.15)$$

It can be seen that the relationship of M_i given above is an extension of the basic notion data = model + noise. Now, using the estimates of the regression coefficient $\hat{\beta}$ and the Breslow estimator Equation 5.12, we obtain the *martingale residual process*:

$$\hat{M}_i(t) = N_i(t) - \int_0^t Y_i(s)e^{\hat{\beta}' x_i(s)}d\hat{H}_0(s). \quad (5.16)$$

In practical settings, we take

$$M_i = \hat{M}_i(\infty) = N_i(t) - \hat{H}_i(\infty).$$

The expected value of the martingale residual is zero, that is, $E(M_i) = 0$, and the sum of the martingale residuals is again zero, $\sum_{i=1}^n M_i = 0$. However, unlike the OLS, ordinary least squares residuals, the martingale residuals can not be used for carrying out goodness-of-fit test.

An alternative to the martingale residual is the *deviance residual* given by

$$d_i = \mathtt{sign}(\hat{M}_i) \times \sqrt{-\hat{M}_i - N_i \log\left(\frac{N_i - \hat{M}_i}{N_i}\right)}, i = 1, \ldots, n. \quad (5.17)$$

The next type of residual we consider is the *score residual*. Now, the score process for the i-th individual is given by

$$U_i(\boldsymbol{\beta}, t) = \int_0^t [\mathbf{x}_i(s) - \bar{\mathbf{x}}(\boldsymbol{\beta}, s)] \, dM_i(s), i = 1, \ldots, n,$$

where

$$\bar{\mathbf{x}}(\boldsymbol{\beta}, s) = \frac{\sum Y_i(s) e^{\boldsymbol{\beta}' \mathbf{x}_i(s)} \mathbf{x}_i(s)}{\sum Y_i(s) e^{\boldsymbol{\beta}' \mathbf{x}_i(s)}}.$$

Note that $U_i(\boldsymbol{\beta}, t)$ is a vector with p number of components, and we can write $U_{il}(\boldsymbol{\beta}, t), l = 1, \ldots, p$. As Therneau and Grambsch argue, the score processes can be viewed as a three-way array with dimension of the subject, covariate, and time. Allowing t_k to denote the time of the k-th event, we can define

$$U_{ilk}(\boldsymbol{\beta}) = \int_{t_{k-1}}^{t_k} [x_{il}(s) - \bar{x}_l(\boldsymbol{\beta}, s)] \, dM_i(s), l = 1, \ldots, p, i = 1, \ldots, n.$$

Now, the *score residual* is defined by $U_{il} = U_{il}(\hat{\boldsymbol{\beta}}, \infty)$, which leads to the score residuals forming an $n \times p$ matrix.

The maximum of the array can be used as a test of the proportional hazards model, see page 85 of Therneau and Grambsch. Further, the score residual are useful for assessing the individual influence on the regression coefficient estimates.

DFFITS and DFBETAS are jackknife residual procedures in the regression analysis. The central idea is the jackknife method. To assess the influence of the i-th observation on an estimate, say $\hat{\theta}$ of θ, we obtain the estimate of θ based on all observations sans the i-th observation and denote it by $\hat{\theta}_{(i)}$ and look at the values of $\hat{\theta} - \hat{\theta}_{(i)}, i = 1, \ldots, n$. An appropriate scaling is also involved. However, we will not go into the technical details of DFBETAS for the relative risk model and simply accept the values provided by R software. For the regression coefficient vector $\boldsymbol{\beta}$, we will have an $n \times p$ matrix, with each row consisting of p elements where a value tells us the influence of the observation on the point estimate of $\beta_k, k = 1, \ldots, p$. A thumb rule in the context of linear regression modeling is that the absolute value of the DFBETAS should not exceed $2/\sqrt{n}$. However, there are no hard and fast rules and one should rather inspect the values before arriving at any conclusion.

A different type of residual for evaluating whether the variables satisfy the assumption of proportionality in the hazards model is obtained by summing over individuals resulting in a process that varies over time, and such residuals are called as the *Schoenfeld residuals*. Since we are summing over the observations, the number of residuals (vector) we get is the number of the events observed in the study. The Schoenfeld residuals at the i-th observed event is given by

$$s_i = \int_{t_{i-1}}^{t_i} \sum_{i'} \left[\mathbf{x}'_i - \bar{\mathbf{x}}(\hat{\boldsymbol{\beta}}, s) \right] d\hat{M}_i(s). \tag{5.18}$$

Here, i' varies over all the individuals while i corresponds to the failure times. At each occurrence of the event, Schoenfeld residuals form a p column vector. We will next look at analysis of the residuals for the PBC dataset.

Example 23 *Residual Analysis of PBC Data Under the Cox Proportional Hazards Model. We will begin with obtaining the martingale and deviance residuals. We extract the residuals for the fitted relative risk model using the* residuals *function with the options specified in the* type *as* "martingale" *and* "deviance" *respectively. An unusual value of the residual indicates the presence of an outlier. The basic summary statistics can be used to get a brief understanding of the residuals. However, we do not have any guidance over the values of the residuals. Nevertheless, we can perform the simple plot of the residuals against the serial number of the observations using* plot.ts *function and it will clearly show if some observations have too high residual values, and whether the fitting of the model through the partial likelihood maximisation has resulted in systematic bias.*

```
> ## Residual Analysis for the Cox PH Regression Model
> residuals(pbc_ph,type="martingale")
      1       2       3       4       5       6       7       8
-0.4139 -0.5130  0.5567 -0.3546 -0.1238  0.7356 -0.1042  0.8454

    305     306     307     308 309      310     311     312
 -0.0855 -0.1188 -0.0398 -0.1013 -0.1999 -0.1143 -0.0578 -0.0345
> residuals(pbc_ph,type="deviance")
      1       2       3       4       5       6       7       8
-0.3676 -1.0129  0.7167 -0.3197 -0.4977  1.0905 -0.4565  1.4295

    305     306     307     308     309     310     311     312
-0.4134 -0.4874 -0.2821 -0.4501 -0.6323 -0.4781 -0.3401 -0.2626
> pdf("../Output/PBC_Martingale_Deviance_Residuals.pdf",
+   height=15,width=10)
> par(mfrow=c(2,1))
> plot.ts(residuals(pbc_ph,type="martingale"),
+        ylab="Martingale Residual",xlab="Observation Number")
> title("RESIDUAL PLOTS")
> plot.ts(residuals(pbc_ph,type="deviance"),
+        ylab="Deviance Residual",xlab="Observation Number")
> dev.off()
```

Figure 5.5 does not show any systematic pattern in the way residuals are associated with the observations. It looks more like a noise process.

The other types of residual—score, Schoenfeld, and DFBETAS—are typically associated with the covariates. The score and DFBETAS residuals have dimensions of the order $n \times p$, while dimension of the Schoenfeld residual is

RESIDUAL PLOTS

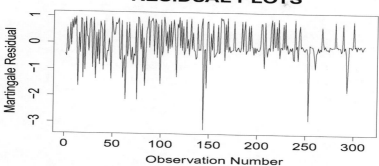

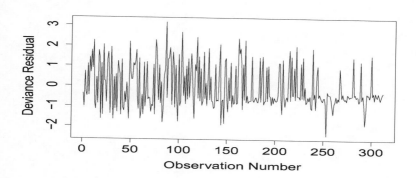

FIGURE 5.5: Martingale and Deviance Residuals for the PBC Data

the number of events observed times the number of covariates. A look at these residuals is in order before we decide which observations need to be removed for causing distortions in the hazards model.

```
> round(residuals(pbc_ph,type="score"),3)
        trt      age   sexf  edema     bili albumin
1     0.183  -0.345 -0.043 -0.131   -1.210   0.122
2     0.249  -2.643 -0.107  0.056    0.641  -0.294
3    -0.186   9.415 -0.412  0.208   -0.487   0.126
4     0.196   1.476 -0.123 -0.008    4.496   0.102
5    -0.066   1.960 -0.023  0.029    0.384  -0.031
6     0.389   9.836  0.085 -0.055   -0.564   0.226

307  -0.020   0.924 -0.007 0.013 0.316   -0.014
308   0.050  -0.749 -0.018 0.032 0.834   -0.039
309  -0.097  -0.888 -0.035 0.068 1.826    0.088
310   0.060  -0.963 -0.019 0.041 0.952   -0.019
311   0.031   0.926 -0.009 0.022 0.477    0.001
```

```
312 -0.016   0.722 -0.006 0.013 0.134   -0.021
> round(residuals(pbc_ph,type="dfbetas"),3)
        [,1]    [,2]    [,3]    [,4]    [,5]    [,6]
1      0.034   0.004 -0.004 -0.026 -0.009   0.012
2      0.040 -0.028 -0.032 -0.012   0.003 -0.068
3     -0.028   0.057 -0.083   0.081 -0.002   0.062
4      0.029   0.019 -0.033 -0.011   0.069   0.022
5     -0.010   0.014 -0.004   0.005   0.006 -0.002
6      0.092   0.117   0.053   0.002 -0.010   0.054

307 -0.003   0.007 -0.001 0.001 0.005 -0.001
308   0.006 -0.007 -0.008 0.003 0.009 -0.007
309 -0.024 -0.011 -0.016 0.023 0.024   0.029
310   0.007 -0.008 -0.008 0.008 0.010 -0.001
311   0.006   0.010 -0.001 0.005 0.005   0.003
312 -0.002   0.005 -0.001 0.002 0.002 -0.003
> round(residuals(pbc_ph,type="schoenfeld"),3)
          trt     age   sexf   edema    bili albumin
41    -0.457   8.482  0.123   0.387   5.883  -0.803
51     0.468  14.549  0.143   0.450   1.548  -0.294
71    -0.510  -3.400  0.150  -0.029   1.219   0.032
77     0.485  -0.937  0.151  -0.029  10.631   0.263
110    0.497  -6.437  0.155   0.470  -8.189   0.629
130    0.499  -9.025  0.155   0.471   6.694  -0.399

3584 -0.510  15.430  0.240  -0.057  -0.482   0.235
3762  0.498   5.151  0.267  -0.064   0.038   0.530
3839  0.545  -7.203  0.211  -0.072   3.373  -0.202
3853  0.561   2.927  0.230  -0.078  -0.346  -0.051
4079 -0.483  -4.383 -0.677  -0.045   0.886   0.347
4191 -0.461  -3.891 -0.688  -0.053   0.183   0.077
```

When we can look at the humongous output of the residual analysis for the covariates, it becomes difficult to generate insights. Toward this, we will choose the score and DFBETAS residuals. We will first plot the score residuals against the covariate values and then impose the the DFBETAS residual values on the same plot and show them in red color. When we see that at any given value of the covariate, the corresponding residual value, be it score or DFBETAS, is either too high or too low, relatively, the observation would then be getting badly influenced by this covariate. Such observations must be presented to domain specialists, clinicians in our case here, and it must then be judiciously marked whether such observation needs to be recorded again, or the values be trimmed. Thus, the data might contain huge noises which can be fixed here. We first obtain the plots, and then have a look at the resultant output.

```
> pbc_score <- as.data.frame(residuals(pbc_ph,type="score"))
> pbc_dfbetas<-as.data.frame(residuals(pbc_ph,type="dfbetas"))
> names(pbc_dfbetas) <- names(pbc_score)
> pdf("../Output/PBC_Score_DFBETAS_Residual.pdf",
+    height=15,width=10)
> par(mfrow=c(3,2))
> plot(pbc2$trt,pbc_score$trt,xlab="Treatment",ylab="Residual")
> par(new = TRUE)
> plot(pbc2$trt,pbc_dfbetas$trt,axes=FALSE,xlab="",
+     ylab="",col="red")
> plot(pbc2$age,pbc_score$age,xlab="Age",ylab="Residual")
> par(new = TRUE)
> plot(pbc2$age,pbc_dfbetas$age,axes=FALSE,xlab="",
+     ylab="",col="red")
> plot(as.numeric(pbc2$sex),pbc_score$sexf,xlab="Sex",
+     ylab="Residual")
> par(new = TRUE)
> plot(as.numeric(pbc2$sex),pbc_dfbetas$sexf,axes=FALSE,
+     xlab="",ylab="",col="red")
> plot(pbc2$edema,pbc_score$edema,
+     xlab="Edema",ylab="Residual")
> par(new = TRUE)
> plot(pbc2$edema,pbc_dfbetas$edema,axes=FALSE,
+     xlab="",ylab="",col="red")
> plot(pbc2$bili,pbc_score$bili,
+     xlab="Bilirubin",ylab="Residual")
> par(new = TRUE)
> plot(pbc2$bili,pbc_dfbetas$bili,axes=FALSE,
+     xlab="",ylab="",col="red")
> plot(pbc2$albumin,pbc_score$albumin,
+     xlab="Albumin",ylab="Residual")
> par(new = TRUE)
> plot(pbc2$albumin,pbc_dfbetas$albumin,axes=FALSE,
+     xlab="",ylab="",col="red")
> dev.off()
```

Result of the above code block is Figure 5.6. Plots corresponding to the treatments gender and edema do not reflect presence of any outliers. However, at the right bottom of the age-residual plot, a value looks unusually large and negative, especially at the average age in that neighborhood. This value needs to be inspected and checked for outliers. The residual plot corresponding to bilirubin shows an outlier at the same right-bottom position. The albumin plot does not show any outlier.

□

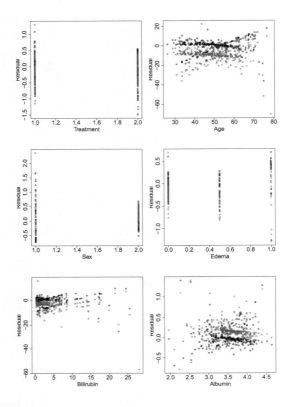

FIGURE 5.6: Score and DFBETAS Residuals for the PBC Data

Residual analysis is an important diagnostic tool in regression analysis. The interpretations and conclusions arrived at after fiting a model without checking validity of the assumptions are prone to be erroneous. Any variant of the proportional hazards model form requires validation methods be derived from scratch.

5.5   Parametric Regression Models

We now consider parametric regression models. In the relative risk model 5.6 $h_i(t|\mathbf{x}_i) = h_0(t)r(\boldsymbol{\beta}, \mathbf{x}_i(t))$, we have two important components. The first is the baseline hazard rate $h_0(t)$ and the second part is the risk term $r(\boldsymbol{\beta}, \mathbf{x}_i(t))$. Since the baseline hazard rate is completely unspecified and any nonnegative function that meets the properties of the hazard rate, the family of functions

$h_0(t)$ satisfying these criteria is very large. In fact, it can be proved mathematically, see Bickel, et al. (1993)[12], that the class of functions meeting the requirement of $h_0(t)$ forms an infinite dimensional space.

The baseline hazard rate can be specified in terms of parametric models. For instance, if we specify that the baseline hazard rate is a constant, say $h_0(t) = k, k > 0$, we are setting up exponential regression models. Similarly, by specifying the baseline hazard rate as one of the parametric lifetime distribution, we can create *parametric regression models*. The parametric regression models are also an important variant of regression model. Borgan (1984)[13] constructed the parametric regression models under the counting process framework to derive and study the MLEs. Section VII.6 of ABGK[7] and Chapter 5 of ABG[2] give detailed exposition on the parametric regression models where the models are of the following form:

$$h_i(t) = Y_i(t)h_0(t,\theta)r(\boldsymbol{\beta}, \mathbf{x}_i(t)), \theta \in \mathcal{R}. \tag{5.19}$$

The baseline hazard rate can be one of the lifetime distributions given in Table 1.1. For instance, by specifying $h_0(t) = \frac{\beta}{\lambda}\left(\frac{t}{\lambda}\right)^{\beta-1}$, we have the Weibull regression model, and taking $h_0(t) = \frac{1}{\Gamma(\beta)\lambda^\beta}\frac{t^{\beta-1}\exp(-t/\lambda)}{1-\Gamma(t/\lambda,\beta)}$ we have the gamma regression model. Note that we have interest in the regression coefficients associated with covariates too, that is, in $\boldsymbol{\beta}$. We will not get into the details of obtaining the regression coefficients $\boldsymbol{\beta}$ or the parameter(s) θ associated with $h_0(t,\theta)$. The interested reader might refer to ABG or ABGK.

The parametric regression models are illustrated with two examples—(i) Weibull regression model for the rat dataset and (ii) loglogistic regression model for analysis for acute myelogenous leukemia data. Parametric regression models are setup using the `survreg` function from the `survival` package.

Example 24 *Weibull Regression Analysis for Rat Data. Here, we have data on $n = 90$ rats. The purpose of the study is to find if the three diet programs have the same effect on the tumor-free period for the rats. The three diets were low-fat, saturated fat, and unsaturated fat. The dataset is provided in the code bundle and the file name is* `Rat.csv`. *Lee and Wang (2003)[71], Table 3.4, contains the data related to this study, and the analysis of Weibull regression model can be obtained on page 270 of the same book.*

Data is drawn in R using the `read.csv` *command. The data variables* `Time` *and* `Delta` *contain time to tumour-free and the corresponding event indicator. The indicator variables* `LOW`, `SAT`, *and* `UNSA` *show whether that diet was given to the rat or not. Obviously because a rat is given one of the three diets, we will not specify all the three variables in setting up the parametric regression model.*

Setting up the survival object of the observed times and the associated indicators with `Surv(Time,Delta)` *and specifying the covariates with* `~SAT+UNSA`, *we setup with Weibull regression model with the option of* `dist="weibull"` *inside the* `survreg` *function.*

```
> # Fitting a Weibull Distribution
> rat <- read.csv("../Data/Rat.csv",header=TRUE)
> str(rat)
'data.frame': 90 obs. of  5 variables:
 $ Time : int   140 177 50 65 86 153 181 191 77 84 ...
 $ Delta: int   1 1 1 1 1 1 1 1 1 1 ...
 $ LOW  : int   1 1 1 1 1 1 1 1 1 1 ...
 $ SAT  : int   0 0 0 0 0 0 0 0 0 0 ...
 $ UNSA : int   0 0 0 0 0 0 0 0 0 0 ...
> rat_weibull <- survreg(Surv(Time,Delta)~SAT+UNSA,data=rat,
+                     dist="weibull")
> summary(rat_weibull)

Call:
survreg(formula = Surv(Time, Delta) ~ SAT + UNSA, data = rat,
    dist = "weibull")
              Value Std. Error     z         p
(Intercept)  5.4004     0.1127 47.93  < 2e-16
SAT         -0.3937     0.1447 -2.72   0.0065
UNSA        -0.7391     0.1395 -5.30 1.2e-07
Log(scale)  -0.8436     0.0996 -8.47  < 2e-16

Scale= 0.43

Weibull distribution
Loglik(model)= -382.9   Loglik(intercept only)= -397.5
Chisq= 29.16 on 2 degrees of freedom, p= 4.7e-07
Number of Newton-Raphson Iterations: 5
n= 90
```

The p-value associated with the overall model fit is p= 4.7e-07, *and this indicates the null hypothesis that the model is insignificant be rejected. In simple words, the fitted model is useful. Further, both the variables* SAT *and* UNSA *are significant as indicated by their respective p-values of* 00065 *and* 1.2e-07, *in simple words, the tumor free time of rats depends on the diet received.*

□

The *loglogistic distribution* is a useful lifetime distribution, and the survival function of a lifetime variable following this distribution is given by

$$S(t) = \frac{1}{1 + \lambda t^{\alpha}}, t > 0, \lambda > 0, \alpha > 0,$$

and the cumulative hazard function is

$$H(t) = \ln\left(1 + \lambda t^{\alpha}\right).$$

We will fit Weibull distribution for two samples first, and then fit the corresponding loglogistic distribution. Using an indicator variable as a covariate, we will then fit the parametric regression model with loglogistic as the probability distribution.

Example 25 *Loglogistic Regression Analysis for Acute Myelogenous Leukemia Data.* *The* alloauto *dataset consists of 90 patients. Here, 101 patients are observed for acute myelogenous leukemia. The recorded times are leukemia-free survival durations, and each patient was seen to have at least one year of leukemia-free duration before remission. The purpose of the study is to compare the efficacy of autologous (auto) versus allogeneic (allo) transplants for acute myelogenous leukemia. The dataset is introduced in Section 1.7 of Klein and Moeschberger (2003)[64].*

The dataset alloauto *is loaded from the Klein and Moeschberger's R package* KMsurv. *We first fit a Weibull distribution for the autologous treatment and obtain the parameters. Next, we repeat the exercise for the allogeneic transplant.*

```
> # Fitting Loglogistic Distribution and comparison with Weibull
> data("alloauto")
> str(alloauto)
'data.frame': 101 obs. of  3 variables:
 $ time : num  0.03 0.493 0.855 1.184 1.283 ...
 $ type : int  1 1 1 1 1 1 1 1 1 1 ...
 $ delta: int  1 1 1 1 1 1 1 1 1 1 ...
> auto_weibull <- survreg(Surv(time,delta)~1,dist = "weibull",
+                         data=alloauto,subset =(type==2))
> summary(auto_weibull)

Call:
survreg(formula = Surv(time, delta) ~ 1, data = alloauto,
    subset = (type == 2), dist = "weibull")
            Value Std. Error    z        p
(Intercept) 3.452      0.218 15.82 <2e-16
Log(scale)  0.105      0.158  0.67   0.51

Scale= 1.11

Weibull distribution
Loglik(model)= -123.4   Loglik(intercept only)= -123.4
Number of Newton-Raphson Iterations: 5
n= 51

> summary(auto_weibull)$var
            (Intercept) Log(scale)
(Intercept) 0.047606472 0.009379547
```

```
Log(scale)   0.009379547 0.024952751
> allo_weibull <- survreg(Surv(time,delta)~1,dist = "weibull",
+                               data=alloauto,subset =(type==1))
> summary(allo_weibull)


Call:
survreg(formula = Surv(time, delta) ~ 1, data = alloauto,
    subset = (type == 1), dist = "weibull")
            Value Std. Error   z       p
(Intercept) 4.254      0.478 8.90 < 2e-16
Log(scale)  0.665      0.189 3.51 0.00044

Scale= 1.94

Weibull distribution
Loglik(model)= -96   Loglik(intercept only)= -96
Number of Newton-Raphson Iterations: 5
n= 50

> summary(allo_weibull)$var
            (Intercept) Log(scale)
(Intercept)   0.2286314 0.04508360
Log(scale)    0.0450836 0.03579631
```

For the autologous transplants, the estimated parameters are $\hat{\mu}_{auto} = 3.452$ and $\hat{\sigma}_{auto} = 1.11$. The original parameters of the Weibull distribution in terms of μ and σ are related by $\lambda = \exp(-\mu/\sigma)$ and $\beta = 1/\sigma$. Similarly, for the allologous translant we have $\hat{\mu}_{allo} = 4.25$ and $\hat{\sigma}_{auto} = 1.94$. The shape parameter estimate for the autologous transplant is $1/1.11 = 0.901$, while that for allologous transplant is $1/1.94 = 0.5155$. This indicates that while exponential distribution might also fit reasonably well for the autologous sample, it does not hold true for the other. We next fit the loglogistic distribution.

We can obtain the variance-covariance matrix for the estimators of the two parameters by running summary(allo_loglogistic)$var.

The specification of "loglogistic" *in the* dist *argument accomplishes the fitting of the loglogistic distribution.*

```
> auto_loglogistic <- survreg(Surv(time,delta)~1,dist =
+                "loglogistic",data=alloauto,subset =(type==2))
> summary(auto_loglogistic)


Call:
survreg(formula = Surv(time, delta) ~ 1, data = alloauto,
    subset = (type == 2), dist = "loglogistic")
            Value Std. Error    z      p
(Intercept) 2.944      0.231 12.77 <2e-16
```

```
Log(scale)  -0.158        0.160 -0.99    0.32

Scale= 0.854


Log logistic distribution
Loglik(model)= -122.2    Loglik(intercept only)= -122.2
Number of Newton-Raphson Iterations: 4
n= 51


> summary(auto_loglogistic)$var
            (Intercept)  Log(scale)
(Intercept) 0.053137067 0.009972345
Log(scale)  0.009972345 0.025656690
> allo_loglogistic <- survreg(Surv(time,delta)~1,dist =
+              "loglogistic", data=alloauto,subset =(type==1))
> summary(allo_loglogistic)

Call:
survreg(formula = Surv(time, delta) ~ 1, data = alloauto,
subset = (type == 1), dist = "loglogistic")
            Value Std. Error   z        p
(Intercept) 3.443      0.476 7.23 4.8e-13
Log(scale)  0.460      0.185 2.49    0.013

Scale= 1.58

Log logistic distribution
Loglik(model)= -94.8   Loglik(intercept only)= -94.8
Number of Newton-Raphson Iterations: 4
n= 50


> summary(allo_loglogistic)$var
            (Intercept)  Log(scale)
(Intercept)  0.22662368 0.03670613
Log(scale)   0.03670613 0.03411400
```

Here, the estimated parameters for the autologous transplants are $\hat{\mu}_{auto} = 2.944$ and $\hat{\sigma}_{auto} = 0.854$, while for the allologous translant $\hat{\mu}_{allo} = 3.443$ and $\hat{\sigma}_{auto} = 1.58$. Whereas the difference between the leukemia-free times under the autologous transplant and the allologous transplant appear significant, $3.443 - 2.944 = 0.499$, we need to answer if it is statistically significant. We fit the parametric regression model using the survreg function and now specify the type of transplant as a covariate to the model.

```
> all_loglogistic <- survreg(Surv(time,delta)~type,dist =
+                         "loglogistic", data=alloauto)
```

```
> summary(all_loglogistic)

Call:
survreg(formula = Surv(time, delta) ~ type, data = alloauto,
    dist = "loglogistic")
                Value Std. Error    z       p
(Intercept)   3.2656     0.7418  4.40  1.1e-05
type         -0.0808     0.4481 -0.18    0.86
Log(scale)    0.1694     0.1213  1.40    0.16

Scale= 1.18

Log logistic distribution
Loglik(model)= -220.2   Loglik(intercept only)= -220.2
Chisq= 0.03 on 1 degrees of freedom, p= 0.86
Number of Newton-Raphson Iterations: 4
n= 101
```

The p-value associated with type *is 0.86 and this difference is insignificant.*
□

The parametric regression models are a useful alternative to the semi-parametric regression models discussed earlier, and also to the nonparametric regression models. The survreg function covers a large family of lifetime distributions. While it is ready to use for six standard distributions, figure them out by running ?survreg at the R console, the reader can custom define the lifetime distribution and use them so long as they satisfy the statistical properties of a lifetime distribution.

5.6 Exercises

Exercise 5.1 *Extend the Koul-Susarla-van Ryzin function* KSvR *to run the bootstrap method to obtain the p-values for testing significance of the covariates.*
□

Exercise 5.2 *Extend the Miller's function* Miller *to run the bootstrap method to obtain the p-values for obtaining the significance of the covariates.*
□

Exercise 5.3 *Regression Analysis for the Veteran's Lung Cancer study. Fit the Cox proportional hazards model, continuation of Exercise 4.3, for the life time of the veterans. Using the dataset* veteran *from the* survival *package, evaluate the impact of other variables, celltype, Karnofsky performance score, Karnofsky performance score, and age of the veteran.*
□

Exercise 5.4 *Perform the residual analysis for the veterans' lifetimes modeled in the previous exercise.* □

Exercise 5.5 *Fit appropriate Cox proportional hazards regression model for the time to event of interest in Example 25 and using the AIC and BIC metrics, infer on better fitting models.* □

5.7 Más Lejos Temas

Sections 5.2–5.5 offer a host of options for setting up the regression models associated with the survival data. While each method serves a purpose or more, the Cox relative risk model has found more acceptance in the medical community. Chapter 6 provides further alternatives with slightly more flexible models. The linear regression methods of Section 5.2 are not widely popular. However, if the solutions under these methods are stable and the predictions can be alligned with the other type of models, the practitioner must not ignore or dump simply because they are not widely used. It will be seen in second part on machine learning methods that any model that is moderately useful must not be discarded.

The Cox model is undoubtedly the preferred model by the practitioners and its theory has also been widely developed. The model has seen lot of extensions and variants. Within the univariate lifetime context, O'Quigley (2008)[89] is a comprehensive account of options: time dependent covariates, non-proportional hazards, partial proportional hazards, estimating equation perspective, appropriate tests based on Brownian motion, changepoint models, etc. Therneau and Grambsch (2000)[116] extend the Cox model in many directions and detail different facets: stratified Cox models, martingale residual analysis, Poisson approach, tests for the proportional hazards assumption, frailty models, etc.

Frailty models is potentially among the most important extensions of the relative risk model. It is similar to the *mixed effects model* in the context of linear regression model. Excess variability is not captured by many models and the search for useful explanatory variables is in vain too. It is in such scenarios that the frailty models provide an useful alternative. O'Quigley (2008) [89], Therneau and Grambsch (2000) [116], and Duchateau and Jannsen (2008) [37] provide the detailed account of frailty models.

Chapter 6

Further Topics in Regression Models

6.1 Introduction

In Chapter 5, we developed multiple ways of modeling the impact of the covariates on the lifetimes in the clinical trials. Direct modeling and hazard rate modeling were introduced. Statistical inference related to the the models were carried out and we also saw how to carry out the residual analysis for the relative risk model. The assumption of proportional hazards is central to the success of the Cox model and it is not always satisfied. If the analyst wants to continue modeling the covariate impact on the hazard rate while being reluctant on the choice of parametric models, the Aalen's additive hazards model comes handy. We will take up this technique in the next section.

Computational power has enabled statisticians to go for techniques which are intensive in the run time. Klein and Andersen (2005)[63] introduced the concept of *pseudovalues* based on the jackknife concept. Using the mean based on the overall observations and again obtaining the same by deleting an observation, pseudovalues are defined as a function of the difference between the two averages. The pseudovalues are not independent of each other and we need to appeal to the broader framework of *Generalized Estimating Equations* to help fit the regression model on the pseudo lifetime values and the covariates. This development will unfold in Section 6.3.

Time dependent covariates are a practical possibility in clinical trials. All the methods discussed hitherto assume that the covariates are static variables throughout the study. Using the same model as earlier, we will now illustrate the modeling of time dependent covariates in Section 6.4.

6.2 Aalen's Additive Regression Model

The Cox proportional hazards model remains the most popular regression technique when it comes to analysis of lifetime data. Though the relative risk model is a broader class than proportional hazards, the assumption of proportional hazards is not satisfied in many scenarios. The parametric regression

DOI: 10.1201/9781003306979-6

models, Section 5.5, might not come to the rescue either and covariates are too important to ignore them in the analysis. Aalen (1980)[1] proposed a complete nonparametric additive regression model on the hazard rate.

We recall the general form of the hazards regression model $h_i(t) = Y_i(t)r(\boldsymbol{\beta}, \mathbf{x}_i(t))$. The specification $r(\boldsymbol{\beta}, \mathbf{x}_i(t)) = h_0(t)e^{\boldsymbol{\beta}' \mathbf{x}_i(t)}$ gives us the proportional hazards regression model. Even as we allow the covariates to vary with time, natural in longitudinal studies, the regression coefficients can be allowed to vary with time. Aalen (1980)[1] proposed an additive hazards regression model where the intensity process of the counting process $N_i(t)$ is given by the following:

$$h_i(t) = Y_i(t)\{\beta_0(t) + \beta_1(t)x_{i1}(t) + \ldots + \beta_p(t)x_{ip}(t)\}, i = 1, \ldots, n. \quad (6.1)$$

The hazard rate model specified by Equation 6.1 is the well-known *Aalen's additive regression model*.

Note that we do not have regression coefficients anymore but regression functions $\beta_0(t), \beta_1(t), \ldots, \beta_p(t)$. Also, as we will later estimate the regression function, there is a possibility that for some t and $\mathbf{x}_i(t)$, the hazard rate $h_i(t)$ might be negative. However, in several applications of Aalen's model made till now, negative estimates of hazard rate has not been a practical problem. ABG, page 155–6, provide many reasons as to why the additive model is worth consideration. Martinussen and Schieke (2006)[76] is also a comprehensive treatise on the topic.

In the absence of covariates, the problem reduces to estimation of the hazard rate. Since estimation of hazard rate is difficult, we, instead, look up to estimating the cumulative hazard function by first obtaining an estimate of the cumulative regression function $B_l(t) = \int_0^t \beta_l(s)ds$. Define the cumulative regression function $\mathbf{B}(t)$ as

$$\mathbf{B}(t) = (B_0(t), B_1(t), \ldots, B_p(t)),$$

where $B_l(t) = \int_0^t \beta_l(s)ds, l = 0, 1, \ldots, p$. In the previous section, the vector of regression coefficients did not include the baseline hazard rate. There ought not to be a confusion as to we have regression function, and thereby we can assume that whenever we have the time dimension, the baseline need not be specified explicitly.

Define the vector of counting processes by $\mathbf{N}(t) = (N_1(t), N_2(t), \ldots, N_n(t))'$, with $N_i(t)$ counting the occurrence of the event by time t for the i-th observation, and the associated martingale by $\mathbf{M}(t) = (M_1(t), M_2(t), \ldots, M_n(t))$. We use $\mathbf{h}(t) = (h_1(t), \ldots, h_n(t))'$ to denote the vector of the intensity processes. We now introduce the regression matrix as $\mathbf{X}(t)$, a matrix of order $n \times (p+1)$ whose i-th row is given by $(Y_i(t), Y_i(t)x_{i1}(t), Y_i(t)x_{i2}(t), \ldots, Y_i(t)x_{ip}(t))$. For the counting process corresponding to the i-th observation, stochastic differential equation is given by

$$
\begin{aligned}
dN_i(t) &= h_i(t)dt + dM_i(t) \\
&= Y_i(t)\beta_0(t) + Y_i(t)\beta_1(t)x_{i1}(t) + \ldots + Y_i(t)\beta_p(t)x_{ip}(t) + dM_i(t).
\end{aligned}
$$

The stochastic differential equation for the n observations can then be written as

$$dN(t) = X(t)dB(t) + dM(t). \tag{6.2}$$

We see that, intuitively, regression function differential $d\mathbf{B}(t)$ can be estimated by the ordinary least squares (OLS) solution as

$$d\hat{\mathbf{B}}(t) = \left(\mathbf{X}'(t)\mathbf{X}(t)\right)^{-1}\mathbf{X}'(t)dN(t).$$

The estimate $d\hat{\mathbf{B}}(t)$ is meaningful only if the inverse $\left(\mathbf{X}'(t)\mathbf{X}(t)\right)^{-1}$ exists. Equivalently, we need $\mathbf{X}(t)$ to be of full rank. Now, we use $J(t)$ to denote that $\mathbf{X}(t)$ has full rank. Next, we denote the *generalized inverse* by $\mathbf{X}^-(t)$ and it is obtained by

$$\mathbf{X}^-(t) = \left(\mathbf{X}'(t)\mathbf{X}(t)\right)^{-1}\mathbf{X}'(t). \tag{6.3}$$

Let $T_1 < T_2 < \ldots$ denote the time points of the event occurrence, as before. We can therefore obtain the estimate of the cumulative regression vector function by accumulating the values of $d\hat{\mathbf{B}}(t)$ at the time points $T_1 < T_2 < \ldots$:

$$\hat{\mathbf{B}}(t) = \int_0^t J(s)\mathbf{X}^-(s)dN(s) = \sum_{T_i \le t} J(T_i)\mathbf{X}^-(T_i)\Delta N(T_i). \tag{6.4}$$

To understand the properties of $\hat{\mathbf{B}}(t)$, we define

$$\mathbf{B}^*(t) = \int_0^t J(s)d\mathbf{B}(s).$$

It is then easy to show that

$$\hat{\mathbf{B}}(t) - \mathbf{B}^*(t) = \int_0^t J(s)\mathbf{X}^-(s)dM(s) \tag{6.5}$$

is a zero-mean martingale, and hence $\hat{\mathbf{B}}(t)$ is an unbiased estimator of $\mathbf{B}^*(t)$. The *predictable variation process* of $\hat{\mathbf{B}} - \mathbf{B}^*$ is then given by

$$\langle \hat{\mathbf{B}} - \mathbf{B}^* \rangle(t) = \int_0^t J(s)\mathbf{X}^-(s) \; \texttt{diag}\{\mathbf{h}(s)ds\}\mathbf{X}^-(s)', \tag{6.6}$$

where $\texttt{diag}\{\mathbf{h}(s)ds\}$ is the diagonal matrix with $\mathbf{h}(s)ds$ as the diagonal vector. An estimator of $\langle \hat{\mathbf{B}} - \mathbf{B}^* \rangle(t)$ is obtained by estimating $\mathbf{h}(s)ds$ with $dN(s)$, and hence covariance matrix estimator is given by

$$\hat{\boldsymbol{\Sigma}}(t) = \sum_{T_i \le t} J(T_i)\mathbf{X}^-(T_i) \; \texttt{diag}\{\Delta N(T_i)\}\mathbf{X}^-(T_i)'. \tag{6.7}$$

By the martingale central limit theorem, it follows that $\sqrt{n}(\hat{\mathbf{B}} - \mathbf{B}^*)$ converges in distribution to a zero mean Gaussian martingale with covariance matrix

$n\langle\hat{\mathbf{B}} - \mathbf{B}^*\rangle$. As with the earlier asymptotics, $\mathbf{B}^*$ and $\mathbf{B}$ are asymptotically equivalent, and hence $\sqrt{n}(\hat{\mathbf{B}} - \mathbf{B})$ converges in distribution to a zero mean Gaussian martingale with same covariance matrix. An $100(1-\alpha)\%$ confidence interval for l-th cumulative regression function $B_l(t)$ is given by

$$\hat{B}_l(t) \pm z_{1-\alpha/2}\sqrt{\hat{\sigma}_{ll}(t)},$$

where $\sqrt{\hat{\sigma}_{ll}(t)}$ is the l-th diagonal element of the estimated covariance matrix. We next apply the development to the PBC dataset.

Example 26 *Modeling the PBC with Additive Hazards Regression.*
In continuation of the analysis of the PBC dataset, we will now model using the Aalen's additive hazards model with the **aalen** *function from the* **timereg** *package. The* **timereg** *package has been created and maintained by Thomas Scheike with contributions from Torben Martinussen. Martinussen and Scheike (2006)[76] is a benchmark reference for the dynamic modeling using the Aalen's model at its core. As in Example 21, we continue to model for the covariates age, treatment, gender, edema, bilirubin, and albumin. The* **summary** *function provides the details.*

```
> # Fitting Aalen Additive Data to PBC
> # Fitting Using the timereg package
> pbc2 <- subset(pbc,select=c(time,status,trt,age,sex,
+                       edema,bili,albumin))
> pbc2 <- na.omit(pbc2)
> pbc_aalen <- aalen(Surv(time,status==2)~trt+age+sex+
+                   edema+bili+albumin,data=pbc2)
> summary(pbc_aalen)
Additive Aalen Model


Test for nonparametric terms
```

	Supremum-test of significance	p-value H_0: B(t)=0
(Intercept)	3.17	0.027
trt	2.03	0.452
age	3.47	0.010
sexf	2.03	0.437
edema	3.82	0.009
bili	5.95	0.000
albumin	4.46	0.001

```
Test for time invariant effects
```

	Kolmogorov-Smirnov test	p-value H_0:constant effect
(Intercept)	4.8200	0.009
trt	0.3150	0.163
age	0.0176	0.250

sexf	0.3660	0.360
edema	0.8200	0.122
bili	0.6220	0.012
albumin	1.0000	0.020

	Cramer von Mises test	p-value H_0:constant effect
(Intercept)	1.74e+04	0.042
trt	4.75e+01	0.374
age	3.46e-01	0.205
sexf	6.01e+01	0.650
edema	8.41e+02	0.116
bili	5.43e+02	0.012
albumin	6.18e+02	0.101

The supremum test tells us that age, edema, bilirubin, and albumin are significant variables. The reader is asked to interpret the results given by the Kolmogorov-Smirnov test and the Cramér-von Mises test.

We obtain a plot of the estimated cumulative regression functions in the following code block:

```
> pdf("../Output/PBC_Aalen_Hazards_Regression_Function.pdf",
+     height=20,width=10)
> par(mfrow=c(2,4))
> plot(pbc_aalen)
> dev.off()
null device
          1
```

The cumulative regression functions $\hat{B}_l(t)$ can be seen varying over time for the six variables of interest along with the intercept function in Figure 6.1. The confidence bands for each of the covariates age, edema, and bilirubin almost entirely lie above the zero line. On the other hand, the confidence band for albumin lies below the zero line. Hence, we can conclude that these covariates have significant influence on time to death under the PBC study.

Aalen's additive hazards regression model can also be fitted using the **aareg** *function from the core* **survival** *package:*

```
> # Fitting using aareg from survival package
> # Alternate way, and nothing more
 pbc_a2 <- aareg(Surv(time,status==2)~trt+age+sex+
+                edema+bili+albumin,data = pbc2)
> round(summary(pbc_a2)$table,4)
            slope    coef  se(coef)       z      p
Intercept  0.0012  0.0122   0.0066  1.8558 0.0635
trt        0.0000  0.0002   0.0009  0.2901 0.7718
age        0.0000  0.0002   0.0000  3.6670 0.0002
```

```
sexf        -0.0001  -0.0013    0.0017  -0.7734 0.4393
edema        0.0020   0.0136    0.0042   3.2311 0.0012
bili         0.0003   0.0013    0.0003   4.6427 0.0000
albumin     -0.0005  -0.0051    0.0015  -3.4507 0.0006
```

The above output is truncated. The results above are consistent with the output obtained in the `timereg` *package.*

□

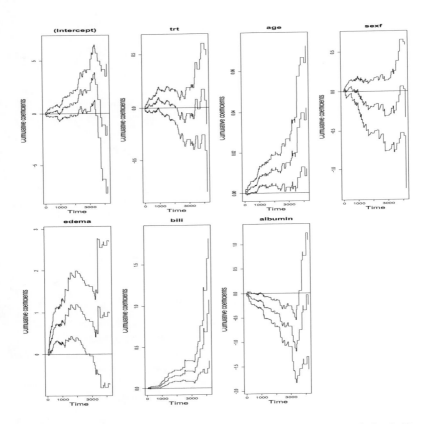

FIGURE 6.1: Cumulative Regression Function Plots in the Aalen's Regression Model

A natural question that crops up while fitting of the additive regression model $h(t) = Y_{(t)} \{\beta_0(t) + \beta_1(t)x_1(t) + \ldots + \beta_p(t)x_p(t)\}$ is that all covariates need not have dynamic impact and that a few of them might have a constant effect. In such cases, we can specify regression coefficients βs instead of regression functions $\beta(t)$. Note that the covariates which are believed not to have dynamic effect might still continue to be time-dependent variables. The models where part covariates have dynamic effect and rest constant in the context of additive models were first proposed and developed by McKeague and Sasieni (1994)[79]. Such models are also *semiparametric models*. Following Martinussen and Schieke (2006)[76], see Chapter 5, we denote by $\mathbf{x}(t) = (x_1(t), \ldots, x_{p_1}(t))$ the (sub)vector of covariates which will have a dynamic impact, and by $\mathbf{z}(t) = (z_1(t), \ldots, z_{p_2}(t))$ the part which will have constant impact. The intensity process of the semiparametric model is given by

$$h_i(t) = Y_i(t)\{\beta_0(t) + \boldsymbol{\beta}'(t)\mathbf{x}_i(t) + \boldsymbol{\gamma}'\mathbf{z}_i(t)\}, i = 1, \ldots, n. \qquad (6.8)$$

We will not get into the mathematical details of estimating $\boldsymbol{\beta}(t)$ or $\boldsymbol{\gamma}$. The same function `aalen` can handle fitting the semiparametric additive regression model. We simply need to change the covariate specification in the model as `const()` for those covariates which have a constant impact.

Example 27 *Fitting a Semi-parametric Additive Regression Model.*
Using `const` *option in the formula specification, we model with the treatment, sex, and edema as variables that have a constant impact while the variables age, bilirubin, and albumin continue to have a dynamic impact. The other part of the program remains same.*

```
> # Fitting the semi-parametric additive Aalen's model
> pbc_a3 <- aalen(Surv(time,status==2)~const(trt)+age+
+                 const(sex)+const(edema)+bili+
+                 albumin,data=pbc2)
> save(pbc_a3,file="../Output/pbc_a3.RData")
> summary(pbc_a3)
Additive Aalen Model


Test for nonparametric terms


Test for non-significant effects
            Supremum-test of significance p-value H_0: B(t)=0
(Intercept)                     2.84                     0.077
age                             3.52                     0.007
bili                            6.29                     0.000
albumin                         4.21                     0.002


Test for time invariant effects
            Kolmogorov-Smirnov test p-value H_0:constant effect
(Intercept)          4.7600                          0.010
age                  0.0101                          0.701
bili                 0.5590                          0.014
albumin              1.0900                          0.010
            Cramer von Mises test p-value H_0:constant effect
(Intercept)       1.90e+04                          0.044
age               9.68e-02                          0.625
```

```
bili                     3.91e+02                           0.009
albumin                  9.63e+02                           0.044
```

```
Parametric terms :
                   Coef.        SE Robust SE
const(trt)       9.58e-06 3.52e-05  3.52e-05
const(sex)f     -7.59e-05 6.82e-05  6.68e-05
const(edema)     5.97e-04 1.87e-04  2.00e-04
                      z    P-val lower2.5% upper97.5%
const(trt)        0.272 0.78500 -5.94e-05   7.86e-05
const(sex)f      -1.140 0.25600 -2.10e-04   5.78e-05
const(edema)      2.980 0.00288  2.30e-04   9.64e-04
  Call:
aalen(formula = Surv(time, status == 2) ~ const(trt) + age +
    const(sex) + const(edema) + bili + albumin, data = pbc2)
```

```
> plot(pbc_a3)
```

Compare the results with earlier fully nonparametric model.   □

Consider a new observation, $\mathbf{x}_0 = (1, x_{01}, \ldots, x_{0p})$. The cumulative hazard function $H(t|\mathbf{x}_0) = \mathbf{x}_0 \mathbf{B}(t)$ can be estimated by

$$\hat{H}(t|\mathbf{x}_0) = \mathbf{x}_0 \hat{\mathbf{B}}(t), \tag{6.9}$$

and with the jump points at the times $T_1 < T_2 < \ldots$, we have

$$\hat{S}(t|\mathbf{x}_0) = \prod_{T_i \leq t} \left(1 - \Delta \hat{H}(T_i|\mathbf{x}_0) \right). \tag{6.10}$$

Example 28 *Survival Function with Aalen's Additive Regression Model. Given the fitted regression model of Example 26, we now undertake the problem of estimating the survival curve for a patient whose history is summarized as follows: the age is 50 years, treatment is D-penicillamine, gender is male, edema is at 0.5 level, while bilirubin and albumin are respectively 10 and 4.*

The fitted **aalen** *object in Example 26 can not be directly used to apply to the new data. We need to add more options of resample and simulation in model specification,* **resample.iid=1,n.sim=0**. *The reasons for their need can be found in the discussion on pages 146–9 of Martinussen and Schieke (2006)[76]. For now, we will simply use the options in* **resample.iid** *and* **n.sim**.

```
> # Prediction with new x0
> pbc_aalen <- aalen(Surv(time,status==2)~trt+age+sex+
+       edema+bili+albumin,resample.iid=1,n.sim=0,data=pbc2)
> xnew <- data.frame(trt=2,age=50,sex="m",edema=0.5,
```

```
+        bili=10,albumin=4)
> levels(xnew$sex) <- levels(pbc2$sex)
> xnew_predict <- predict(pbc_aalen,newdata=xnew,uniform=0)
> plot(xnew_predict)
```

We need to explicitly specify that though we are allowing only male as the gender in the new unknown case with `sex="m"`, *the levels of the factor must be aligned with the* `sex` *variable in the overall dataset. Otherwise, we will run into execution error. The result of the output is Figure 6.2.*

□

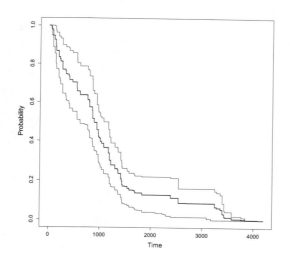

FIGURE 6.2: Interval Estimate of Survival Function with Aalen's Model

Aalen's additive hazards model has considerable flexibility for modeling survival data. We can use it as a semiparametric model too. ABG provide a host of other applications of the additive hazards model. This model is especially useful when the proportional hazards model assumption is not satisfied.

6.3 Regression Based on Pseudo-observations

Chapter 5 introduced regression models for survival data. Though direct modeling of the covariate impact on the survival times is appealing, the incomplete form because of censoring and other issues makes it difficult for carrying out proper statistial inference. The relative risk model and the additive regression model are kind of tailor made methods for censored data, especially the martingale structure coming to the rescue where inference is concerned. The appeal of direct modeling refuses to die down which forces us to look for modeling in some or other form. Andersen, et al. (2003)[9], Andersen, et al. (2004)[8], and Klein and Andersen (2005)[63] provided the breakthrough in the form of jackknife adaptation where pseudo-observations are generated and then using an extended regression framework, the impact of the covariates is modeled on the pseudo-observations. We will follow the development in Andersen, et al. (2004)[8].

Let $T_1, T_2, \ldots, T_n$ be an IID sample of size n from a common distribution $S = 1 - F$. Let θ denote the expected value of $g(T)$, where g is a continuous function:

$$\theta = E(g(T_i)). \tag{6.11}$$

We assume the existence of an unbiased estimator, or an approximately unbiased estimator, of θ in the form of $\widehat{\theta}$, that is, $E(\widehat{\theta}) = \theta$. Let $\mathbf{x}_i, i = 1, \ldots, n$ be IID covariates such that

$$\theta_i = E(g(T_i)|\mathbf{x}_i).$$

Now, we obtain the so-called *leave-one-out* estimator of θ in $\widehat{\theta^{-i}}$ which is based on all observations $T_j, j \neq i$. The i-th *pseudo-observation* is then given by

$$\widehat{\theta}_i = n\widehat{\theta} - (n-1)\widehat{\theta^{-i}}, i = 1, \ldots, n. \tag{6.12}$$

When dealing with censored data, we are often interested in estimating the average lifetime. With the right-censored date, the Kaplan-Meier estimator $\widehat{S}(t)$ has many desirable properties and the area under the survival curve is the mean of the distribution F, that is,

$$\widehat{\mu}(\tau) = \int_0^\tau \widehat{S(s)}ds.$$

Here, τ is the maximum time up to which $S(t) > 0$. In practical settings, we take τ as the maximum of the observed values. To setup the pseudo-observations as required in Equation 6.12, we can take $\widehat{\theta}$ as $\widehat{\mu}(\tau)$ and $\widehat{\theta^{-i}}$ as $\widehat{\mu^{-i}}(\tau)$ that is based on all observations sans the i-th one.

The influence of the covariate $\mathbf{x}_i$ on the i-th observation is then modeled by specifying a link function and the *generalized linear model* as follows:

$$g(\theta_i) = \boldsymbol{\beta}'\mathbf{x}_i, i = 1, \ldots, n. \tag{6.13}$$

The unit link function $g(v) = v$ is useful when we are dealing with the lifetime variable. We simply take $g(\theta_i) = \widehat{\mu^{-i}}(\tau)$ here. Note that the values $\widehat{\mu^{-i}}(\tau)$ are not independent observations, and hence the usual solutions do not come to the rescue. To estimate β in Equation 6.13, we need *generalized estimating equations*, GEE, given by

$$U(\boldsymbol{\beta}) = \sum_{i=1}^{n} U_i(\boldsymbol{\beta}) = \sum_{i=1}^{n} \left(\frac{\delta}{\delta\boldsymbol{\beta}} g^{-1}(\boldsymbol{\beta}'\mathbf{x}_i) \mathbf{V}_i^{-1} (\widehat{\theta}_i - g^{-1}(\boldsymbol{\beta}'\mathbf{x}_i)) \right) = 0,$$

where $\mathbf{V}_i$ is a working covariance matrix for $\widehat{\theta}_i$. Here, θ is a scalar quantity and $\mathbf{V}_i$ is a working variance of $\widehat{\theta}_i$.

We will now apply the pseudo-observations technique to the Bone-marrow transplant dataset. To obtain the pseudo-observatons, we will use the `pseudo` R package. The pseudo-observations $\widehat{\mu^{-i}}(\tau)$ are used in place of the censored observations while the complete observations are left untouched. However, the entire set of observations can not be pretended to be mutually independent and we use the GEE framework for analysis.

Example 29 *Regression Analysis with Pseudo-observations. The bone-marrow transplant dataset is a benchmark dataset and we use it from the* KMsurv *package. The initials KM here are for JP Klein and ML Moeschberger, and the datasets are drawn from their popular 2003 book.*

Before embarking and deploying the new technique of the section, we first carry out the analysis with the Cox proportional hazards model. We will model for the disease-free survival time, t2, *as a function of the numerous covariates. The associated event time indicators are provided in* d3. *The covariates consist of disease group in* group, *Acute GVHD Indicator, Chronic GVHD Indicator, Platelet Recovery Indicator, Patient Age In Years, Donor Age In Years, Patient Sex, Donor Sex, Patient CMV Status, Donor CMV Status, Waiting Time to Transplant In Days, FAB Grade, and hospital indicator. Using the* coxph *function, we fit the proportional hazards regression model.*

```
> # Pseudo Observations and Regression
> # bmt data set is from KMsurv package
> data(bmt)
> leukemia_coxph <- coxph(Surv(t2,d3==1)~group+da+dc+dp+z1+z2+z3+
+                         z4+z5+z6+z7+z8+z9,data=bmt)
> summary(leukemia_coxph)
Call:
coxph(formula = Surv(t2, d3 == 1) ~ group + da + dc + dp + z1 +
    z2 + z3 + z4 + z5 + z6 + z7 + z8 + z9, data = bmt)

  n= 137, number of events= 83

          coef exp(coef)  se(coef)      z Pr(>|z|)
group -0.002208  0.997794  0.195485  -0.01    0.991
da     0.238379  1.269190  0.325631   0.73    0.464
dc    -1.102480  0.332047  0.259272  -4.25  2.1e-05 ***
```

```
dp    -1.683457   0.185731   0.332766  -5.06   4.2e-07 ***
z1     0.016423   1.016558   0.020863   0.79   0.431
z2     0.013383   1.013473   0.019839   0.67   0.500
z3    -0.237607   0.788513   0.254287  -0.93   0.350
z4     0.082076   1.085539   0.255756   0.32   0.748
z5     0.188154   1.207020   0.255195   0.74   0.461
z6    -0.274885   0.759659   0.256935  -1.07   0.285
z7     0.000373   1.000373   0.000334   1.12   0.264
z8     0.557258   1.745878   0.274681   2.03   0.042 *
z9    -0.283236   0.753342   0.113259  -2.50   0.012 *
---
Signif. codes:
0 '***' 0.001 '**' 0.01 '*' 0.05 '.' 0.1 ' ' 1


Concordance= 0.722   (se = 0.028 )
Likelihood ratio test= 57.2  on 13 df,   p=2e-07
Wald test           = 58.1  on 13 df,   p=1e-07
Score (logrank) test = 71.6  on 13 df,   p=4e-10
```

Though the model turns out to be significant and we comprehensively conclud from the three types of chi-square tests that the regression coefficient vector is significantly different from being null, only four out of thirteen variables are significant.

Using the **pseudomean** *function from the* **pseudo** *package, we obtain the pseudo-observations which are stored in* **dfs_pm** *vector. Tempting as it might be to use the* **lm** *function to model impact of covariates on the pseudo-observations, we refrain from that as the observations are not independent of each other. Using the* **gee** *function from the package of same name, we now model the impact of the thirteen variables on the pseudo-observations. The specification of identification number and family of probability distribution is essential in fitting the GEE.*

```
> dfs_pm <- pseudomean(time=bmt$t2,event=bmt$d3,tmax=2000)
> leukemia_gee <- gee(dfs_pm~group+da+dc+dp+z1+z2+z3+
+                      z4+z5+z6+z7+z8+z9,data=bmt,id=1:137,
+                      family="gaussian")
Beginning Cgee S-function, @(#) geeformula.q 4.13 98/01/27
running glm to get initial regression estimate
(Intercept)        group           da           dc
   138.1517      77.6227     -77.3897     416.0073
        dp           z1           z2           z3
   745.3275     -13.4575       1.8455     122.7410
        z4           z5           z6           z7
   -61.7616     -63.9517      85.3973      -0.1133
        z8           z9
  -465.0312     160.8249
```

```
> summary(leukemia_gee)
```

GEE: GENERALIZED LINEAR MODELS FOR DEPENDENT DATA
gee S-function, version 4.13 modified 98/01/27 (1998)

Model:
 Link: Identity
 Variance to Mean Relation: Gaussian
 Correlation Structure: Independent

Call:
```
gee(formula = dfs_pm ~ group + da + dc + dp + z1 + z2 + z3 +
    z4 + z5 + z6 + z7 + z8 + z9, id = 1:137,
    data = bmt, family = "gaussian")
```

Summary of Residuals:
Min	1Q	Median	3Q	Max
-1247.6	-663.1	-131.1	658.1	1548.1

Coefficients:

	Estimate	Naive S.E.	Naive z	Robust S.E.
(Intercept)	138.1517	376.0617	0.3674	338.4796
group	77.6227	108.2107	0.7173	104.6596
da	-77.3897	181.8924	-0.4255	169.0107
dc	416.0073	143.7033	2.8949	138.7252
dp	745.3275	221.6858	3.3621	165.3731
z1	-13.4575	12.0876	-1.1133	10.4537
z2	1.8455	10.6482	0.1733	10.3261
z3	122.7410	142.6823	0.8602	132.3201
z4	-61.7616	146.8547	-0.4206	139.5403
z5	-63.9517	150.1813	-0.4258	147.5843
z6	85.3973	151.0938	0.5652	144.6430
z7	-0.1133	0.1982	-0.5715	0.1906
z8	-465.0312	172.5830	-2.6945	168.8017
z9	160.8249	61.1047	2.6320	57.6909

Estimated Scale Parameter: 620708
Number of Iterations: 1

Working Correlation
```
      [,1]
[1,]    1
```

The summary of the GEE object gave us the point estimate, the standard error, the naive z scores, and a robust estimate of the standard error. However,

it does not give us the p-values. It is easy to obtain them using the pnorm
function which is done next:

```
> 1-pnorm(abs(tt$coefficients[,1]/tt$coefficients[,2]))
(Intercept)        group           da           dc
     0.3567       0.2366       0.3352       0.0019
         dp           z1           z2           z3
     0.0004       0.1328       0.4312       0.1948
         z4           z5           z6           z7
     0.3370       0.3351       0.2860       0.2838
         z8           z9
     0.0035       0.0042
```

*Again, the same set of covariates turn out significant as with the Cox propor-
tional hazards model. However, the practitioner will find it easier to interpret
the regression coefficients in the GEE over the Cox proportional hazards model.*
□

Pseudo-observations are useful as a broader framework too. For instance,
we need not restrict ourselves to the GEE analysis. In the machine learning
framework, Part II, many methods are not tailored to handle censored data.
However, the machine learning methods can be applied to numeric variables
such as pseudo observations.

6.4 Modeling with Time-Dependent Covariates

The independent covariate vector has been written in many places as a variable
that is function of time, $\mathbf{x}(t)$. However, the variables have remained static even
as their impact varies dynamically, as in Aalens additive regression model. In
many clinical trials, the parameters of some covariates change over different
recordings. For instance, the sugar level is a variable that takes different values
over time. Weight is also another example. The mathematics behind fitting the
models does not change. However, we need to suitably prepare the dataset for
proper implementation. The time dependent covariate modeling will be dealt
with here.

A flat row wise dataset can be transformed to the required format using
tmerge function from the survival package.

Example 30 *Cox and Aalen Hazards Regression with Time Dependent Covariates. Here, we have data from a placebo controlled trial of
gamma interferon in chronic granulotomous disease (CGD). The data is on
the time to serious infections observed through end of study for each patient,
and thereby multiple infections are recorded. The explanatory variables recoded*

on the patient are the following. The treatment is either placebo or gamma interferon. Patient characteristics of age, gender, height, and weight are noted at the beginning of the study. Besides these variable, the inheritance pattern, use of steroids at the time of entry, use of prophylactic antibiotics, and a categorization of the hospitilizatin centers is also noted. The data is available in a flat format dataset `cgd0` and in a ready for analysis in `cgd` dataset. The `id` gives the information that multiple rows of information belong to the same patient. A simple R check confirms the data arrangement.

```
> # Basic Testing
> dim(cgd)
[1] 203  16
> length(unique(cgd$id))
[1] 128
> dim(cgd0)
[1] 128  20
> names(cgd)
 [1] "id"       "center"    "weight"
 [9] "inherit"  "steroids"  "status"
> names(cgd0)
 [1] "id"       "center"    "weight"
 [9] "inherit"  "steroids"  "etime3"
[17] "etime4"   "etime5"    "etime7"
```

While there are instances where a patient has only one row in the data file, in some cases a patient is given the treatment, placebo or gamma interferon, up to seven times. Here, we have 128 patients and the multiple rows means we have 203 rows. It is not to be understood that we have 203 different patients. Using the `cluster` option in R, we specify the model functions, `coxph`, `aalen`, and `aareg`, that the covariate data of the patients marked by their `id` varies over time. Thus, the modeling occurs based on 128 patients and not 203 independent rows. We will begin the analysis with the proportional hazards model.

```
> # The Chronic Granulotomous Disease Study
> cgd_ph <- coxph(Surv(tstart,tstop,status==1)~treat+sex+age+
+                 inherit+steroids+cluster(id),data = cgd)
> summary(cgd_ph)
Call:
coxph(formula = Surv(tstart, tstop, status == 1) ~ treat + sex +
    age + inherit + steroids, data = cgd, cluster = id)

  n= 203, number of events= 76

               coef exp(coef) se(coef) robust se
treatrIFN-g  -1.0954    0.3344   0.2638    0.3155
sexfemale    -0.6766    0.5084   0.3904    0.4653
```

```
age                  -0.0394   0.9614   0.0138   0.0149
inheritautosomal      0.6428   1.9018   0.2765   0.3985
steroids              1.4036   4.0697   0.5566   0.6918
                        z Pr(>|z|)
treatrIFN-g          -3.47  0.00052 ***
sexfemale            -1.45  0.14590
age                  -2.64  0.00818 **
inheritautosomal      1.61  0.10673
steroids              2.03  0.04247 *
---
Signif. codes:
0 '***' 0.001 '**' 0.01 '*' 0.05 '.' 0.1 ' ' 1

                 exp(coef) exp(-coef) lower .95 upper .95
treatrIFN-g          0.334      2.990     0.180     0.621
sexfemale            0.508      1.967     0.204     1.265
age                  0.961      1.040     0.934     0.990
inheritautosomal     1.902      0.526     0.871     4.153
steroids             4.070      0.246     1.049    15.792


Concordance= 0.693  (se = 0.041 )
Likelihood ratio test= 34.4  on 5 df,    p=2e-06
Wald test             = 21.7  on 5 df,    p=6e-04
Score (logrank) test = 32.8  on 5 df,    p=4e-06,
              Robust = 12  p=0.03
```

> (Note: the likelihood ratio and score tests
> assume independence of observations within a cluster,
> the Wald and robust score tests do not).

The note above emphasizes that unlike Wald and robust score tests, the likelihood ratio and score tests assume independence. The treatment, age, and use of steroids (before the beginning of the study) are significant in explaining the time to infection for the patients. We next use the Aalen's additive regression model by uing the `cluster` *option.*

```
> cgd_aalen <- aareg(Surv(tstart,tstop,status==1)~treat+sex+age+
+                    inherit+steroids+cluster(id),data = cgd)
> summary(cgd_aalen)
                   slope    coef se(coef) robust se      z      p
Intercept         0.0052  0.0205   0.0039    0.0046  4.501 0.0000
treatrIFN-g      -0.0024 -0.0094   0.0023    0.0029 -3.204 0.0014
sexfemale        -0.0012 -0.0063   0.0040    0.0045 -1.413 0.1576
age              -0.0001 -0.0003   0.0001    0.0001 -2.639 0.0083
inheritautosomal  0.0018  0.0071   0.0041    0.0036  1.988 0.0468
steroids          0.0051  0.0152   0.0102    0.0104  1.468 0.1422
cluster(id)       0.0000 -0.0001   0.0000    0.0000 -1.909 0.0562
```

```
$test
[1] "aalen"


$test.statistic
        Intercept      treatrIFN-g      sexfemale
           19.568          -17.467         -4.221
             age inheritautosomal        steroids
         -199.806            7.078           2.584
        cluster(id)
         -500.690


attr(,"class")
[1] "summary.aareg"
```

We leave the interpretation of the results to the reader. □

In clinical trials, the covariates are more often time-dependent and thus it is important to be able to model for such feature of the data.

6.5 Exercises

Exercise 6.1 *Fit Aalen's additive regression model for the Veteran's life time as extension of the analysis done in Exercises 4.3 and 5.3.* □

Exercise 6.2 *For the* xnew *patient history of Example 28, fit a Cox proportional hazards model with the same set of covariates as in model* pbc_aalen. *Using the fitted Cox model, proceed to complete the survival function plot for* xnew. □

Exercise 6.3 *Create the pseudo-observations for the PBC dataset, perform the regression analysis along the lines of Example 29.* □

Exercise 6.4 *Extract the residuals of the fitted GEE model* leukemia_gee. *Plot the residuals against their indexes, or use* plot.ts. *Check if you find systematic patterns in the residual plot.* □

6.6 Más Lejos Temas

Aalen's additive regression model is a useful alternative to the widely used Cox proportional hazards model. The form of the model introduced in

Section 6.2 is flexible because we are now dealing with regression functions and not regression coefficients. Aalen (1980)[1] has seen several modifications and extensions and the comprehensive text dedicated to these is Martinussen and Scheike (2006)[76] which covers asymptotics, goodness-of-fit, cluster, the multiplicative-additive model, marked point process, etc. Martinussen and Scheike (2007)[77] develop the change point variant of the Aalen's additive regression model. In the next chapter, we will see another variant of the Aalen's additive model where the covariates have fixed effect and are not functions of time. This makes it convenient to compare it with the conventional Cox model.

The pseudo-observations method is essentially delete-one-jackknife technique. The pseudo-observations help in creating a host of flexible models. We will see later in the text that this technique can be used to create more interpretable survival trees. These observations can also be used in the domain of neural networks.

Chapter 7

Model Selection

7.1 Introduction

In all the regression methods discussed thus far, we have either given the features to be included in the model, or did nothing over the insignificant variables. There is a two-fold problem with this approach. In general, it is not possible to say which of the covariates will not be significant. Keeping insignificant variables in the model might mislead a few signs, besides blocking the entry of a few significant variables. Model selection methods are well developed in the linear models theory and the core ideas of forward, backward, and stepwise selection continue to be useful. With likelihood inference leading the way in most streams of the subject, the model selection method has swiftly moved from the p-values to information criteria.

Akaike information criterion (AIC) and Bayesian information criterion (BIC) are two methods based on the information contained in the likelihood functions. It is more robust over the p-values associated with the individual variables. In the context of the Cox proportional hazards model, AIC and BIC are based on the partial likelihood function and thereby the focus is on the influence of the covariates. The baseline is completely ignored while carrying out the model selection. This remark also applies to the parametric regression models based on the important lifetime distributions where the modeling is driven by the hazard rate. Thus, there is need to extend the model selection criteria wherein the failure times also have a say in the process. Nils Lid Hjort and his team developed a new criterion through a series of papers which does this extension. The criterion invented is called as *Focused Information Criterion*, FIC, discussed in Section 7.3.

Model selection can also be carried out by an another mechanism and it is called as *penalization*. Basically, the idea is to add a penalty as the number of parameters increases. We have two choices of penalization—the ℓ_1 penalty, and the ℓ_2 penalty. Computationally, the ℓ_2 is simpler to implement, and in the context of the multiple linear regression, it is equivalent to estimating the coefficients based on the principal components. Sans the intercept term, the penalization methods restrict the overall magnitude of the coefficients. Since the magnitude is constrained, the method is sometimes called as *shrinkage method*. The ℓ_1 penalty is often called as *Least Absolute Shrinkage and*

Selection Operator, or simply LASSO. The abbreviation is so widely in use that practitioners often forget the complete name. The reason for further calling the ℓ_1 penalty as selection operator is that in some cases, the coefficients are returned with an estimated value of 0 which is as good as eliminating the covariate and thereby the variables with nonzero estimated values are selected. These developments for Section 7.4.

Recollect that in discussion on the Aalens additive risk model in Chapter 6, we do not have estimates of the coefficients. Rather, we obtain the estimates of cumulative regression functions. We are not aware of model selection in this setup of the Aalen's additive model. For the semiparametric version of the model where only the baseline depends on time and the influence of other covariates is only through a fixed unknown coefficient. Model selection is possible in this setting and we deal with it in section 7.5.

7.2   Model Selection with AIC and BIC

The Akaike information criterion (AIC) and the Bayesian information criterion (BIC) are important tools in model selection. In Section 2.2, given the estimate $\hat{\theta}$ and the data, we had introduced AIC as

$$AIC = -2\log L(\hat{\theta}, \texttt{data}) + 2k, \qquad (7.1)$$

where k is the number of fitted parameters. We had illustrated the computation in Example 2 and asked the reader to fit a Weibull distribution for the same data in Exercise 2.3 and infer which is a better fit. The abbreviation AIC use is so widespread that some users are unaware of the full form.

The genesis of AIC is in information theory and it was invented by Hirotugu Akaike, Akaike (1973)[5]. A statistical model is used to represent the process that is believed to generate the data. There will be loss of information in the representation and the AIC tries to capture the relative loss of information. Thus, the lesser the value of AIC of a model, the more parsimonious it is for applications. By adding the number of parameters, $2k$ in Equation 7.1, we are penalizing the model where more parameters are used. Thus, if the number of parameters increases and the new variables are useful in explaining the variance of the lifetimes, it is vital that it results in significant decrease in the loss of information as captured by $-2\log L(\hat{\theta}, \texttt{data})$.

Bayesian information criterion (BIC) uses a different penalty and it is given by

$$AIC = -2\log L(\hat{\theta}, \texttt{data}) + \ln(n)k, \qquad (7.2)$$

where n is the number of observations used to create the model. The BIC is sometimes also called as Schwarz information criterion after it was first put forth in Schwarz (1978)[101]. Often, BIC is not provided in many software implementations. Unlike the standard Bayesian analysis, here we do not have prior specification. However, it is implicitly used here and if there was no prior information being used, the term Bayesian would not be used at all. Where the BIC value is not given by the software and only AIC is available, it is not really difficult arithmetic to first substract $2k$ from the AIC value and then add $\ln(n)k$ to obtain the BIC.

We will next look at some instances where we have created multiple models for the same data set and using the AIC decide which fits best among the models employed.

Example 31 *Extracting AIC and BIC Values for Fitted Models. We have fitted many parametric regression models in Section 5.5. Weibull and log-logistic regression models were fitted on the time to relapse of leukemia. The data was obtained from Klein and Moeschberger (2003)[64]. The models fitted in Chapter 5 are loaded here. For the Weibull and the log-logistic fitted regression models, we readily obtain the AIC and BIC using the same named functions.*

```
> list.files("../Data/")
[1] "allo_loglogistic.RData" "allo_weibull.RData"
[3] "cgd_aalen.RData"        "cgd_ph.RData"
[5] "leukemia_coxph.RData"
> load("../Data/allo_weibull.RData")
> load("../Data/allo_loglogistic.RData")
> summary(allo_weibull)
> summary(allo_loglogistic)
> AIC(allo_weibull)
[1] 196
> AIC(allo_loglogistic)
[1] 193.7
> BIC(allo_weibull)
[1] 199.8
> BIC(allo_loglogistic)
[1] 197.5
```

Which of the models allo_weibull *and* allo_loglogistic *is better? We are inclined toward the log-logistic regression model because of the lower AIC and the lower BIC.*

□

In Example 26, treatment and gender were insignificant variables. It means that even if these variables are present in the model, they have no influence on the lifetime of the patient. Similarly, in Example 29, the variables acute GVHD Indicator da and z1-z7 variables are all insigficant. We need a proper variable selection method. We will next consider how to carry out model selection for the relative risk model.

Model selection in the linear regression analysis is driven by the F-statistic and the associated p-value of the covariate effect. The F-statistic can be substituted by either AIC or BIC. This is a positive development in a certain sense because the AIC/BIC are driven by likelihood function. In general, there are three ways of carrying out the model selection using the evaluation criteria. The three methods are (i) backward selection, (ii) forward selection, and (iii) stepwise selection.

In the backward selection method, we begin with a full model. In a full model, we select all possible/available covariates, p here, and setup the model and note the AIC value. Recollect that we need a model which has the least AIC value. Thus, if a variable is removed, we expect a decrease in the AIC value. In the next step, p models are created by dropping one variable at a time and the variable which leads to maximum decrease in the AIC value is eliminated. The search continues, dropping variables, until removing of variables no longer improves the model AIC. The forward selection method begins with a null model, or zero covariates. In the next step, we select the variable which leads to maximum improvisation in the AIC and add it to the model. The search continues until the addition of the variable no longer improves the model performance. The stepwise or bi-directional search at every stage eliminates the most insignificant variable while also trying to find a variable which improves the most. It will stop when neither a variable can be eliminated nor one added. The `step` function works in the R software. Note that the selection method driven by the AIC value is based on the partial likelihood, that is, the baseline hazard $h_0(t)$ is completely ignored.

A few points are important to record here. The backward, forward, and bi-directional selections need not lead to the same final model. It is possible in the bi-directional search that a variable gets dropped at some stage and later gets back into the model. This is no indication of fallacy of the procedure. Also, the search is not exhaustive in nature. All possible models are not considered in any of the three methods. The reader might wonder why does one have to restrict the search. An answer is the quest of all possible model selection is a **nondeterministic polynomial time** NP-hard problem. The total number of possible models is 2^p models, and it is common in clinical trials that p will be in hundreds, and sometimes thousands. Thus, we have that need of efficient model selection procedures.

We will work with bi-directional model selection for the leukemia problem encountered earlier in Example 29.

Example 32 *Model Selection with Cox Proportional Hazards Model Based on AIC*. *In the leukemia study, we have 13 variables and thus the number of all possible models is $2^13 = 8192$. We will first load the full model that was saved as* `leukemia_coxph.RData` *in Chapter 6. Using the* `step` *function with the option* `direction = "both"`, *we carry out the bi-directional search. The other two options for* `direction` *are* `forward` *and* `backward`. *Let us see the action now.*

The total output of the step *function will run into pages and we will use it partially to explain the action.*

```
> # Model Selection with Cox Proportional Hazards Model
> load("../Data/leukemia_coxph.RData")
> summary(leukemia_coxph)
> data(bmt)
> leukemia_both_AIC <- step(leukemia_coxph,direction = "both")
Start:  AIC=715.3
Surv(t2, d3 == 1) ~ group + da + dc + dp + z1 + z2 + z3 + z4 +
    z5 + z6 + z7 + z8 + z9

          Df AIC
- group   1  713
- z4      1  713
- z2      1  714
- da      1  714
- z5      1  714
- z1      1  714
- z3      1  714
- z7      1  714
- z6      1  714
<none>       715
- z8      1  717
- z9      1  707
- dc      1  718
- dp      1  724

Step:  AIC=713.3
```

Here, the software has found that dropping group *will give the best model with least AIC value at* 713.3. *The* − *sign reflects deletion while a* + *sign means addition as seen in next part of the output.*

```
Surv(t2, d3 == 1) ~ da + dc + dp + z1 + z2 + z3 + z4 + z5 + z6 +
    z7 + z8 + z9

          Df AIC
- z4      1  711
- z2      1  712
- z5      1  712
- da      1  712
- z1      1  712
- z3      1  712
- z7      1  712
- z6      1  712
```

```
<none>      713
+ group   1 715
- z8      1 716
- z9      1 718
- dc      1 731
- dp      1 733
```

Step: AIC=711.4

Now, we drop the variable z4, *the AIC value will be* 711.4 *while bringing back* group *would shoot up the AIC value. Hence, we simply drop the* z4 *variable. Similarly, in the next few steps variables will be added/dropped in the following order: –* z2, *–* z5, *–* da, *–* z6, *–* z3, *and –* z7. *The AIC value at this stage is* 702.8. *Adding or dropping any variable leads to worsening the model as seen below:*

```
Step:  AIC=702.8
Surv(t2, d3 == 1) ~ dc + dp + z1 + z8 + z9

        Df AIC
<none>      703
+ z7      1 704
+ da      1 704
+ z2      1 704
+ z6      1 704
+ z3      1 705
+ z5      1 705
+ group   1 705
+ z4      1 705
- z1      1 705
- z8      1 707
- z9      1 707
- dc      1 718
- dp      1 724
```

The final model, deleting 8 out of the 12, is the following:

```
> summary(leukemia_both_AIC)
Call:
coxph(formula = Surv(t2, d3 == 1) ~ dc + dp + z1 + z8 + z9,
data = bmt)
  n= 137, number of events= 83

     coef exp(coef) se(coef)      z Pr(>|z|)
dc -0.973    0.378    0.240  -4.05  5.1e-05 ***
dp -1.688    0.185    0.302  -5.58  2.4e-08 ***
```

```
z1   0.026      1.026      0.012  2.16   0.0307 *
z8   0.587      1.798      0.228  2.58   0.0099 **
z9  -0.269      0.764      0.108 -2.49   0.0128 *
---
Signif. codes:  0 '***' 0.001 '**' 0.01 '*' 0.05 '.' 0.1 ' ' 1

     exp(coef) exp(-coef) lower .95 upper .95
dc     0.378     2.646      0.236     0.605
dp     0.185     5.407      0.102     0.334
z1     1.026     0.974      1.002     1.051
z8     1.799     0.556      1.151     2.809
z9     0.764     1.308      0.619     0.945


Concordance= 0.718  (se = 0.028 )
Likelihood ratio test= 53.8  on 5 df,   p=2e-10
Wald test            = 56    on 5 df,   p=8e-11
Score (logrank) test = 68.6  on 5 df,   p=2e-13
```

To setup the forward and backward selection methods for the leukemia dataset, we need to drop the extraneous variables in the bmt *dataset, particularly the columns labeled by t1,d1,d2,ta,tc,tp, and z10. The* null *and* full *models are setup using the* formula *with the respective specifications* Surv(time,indicator)~1 *and* Surv(time,indicator)~. *. The options of* backward *and* forward *are next used to run the forward and backward selection techniques.*

```
> bmt2 <- subset(bmt,select = -c(t1,d1,d2,ta,tc,tp,z10))
> null <-coxph(Surv(t2,d3==1)~1,bmt2)
> full <-coxph(Surv(t2,d3==1)~.,bmt2)
> leukemia_back_AIC <- step(full,data=bmt2,
+     direction = "backward")
Start:  AIC=715.3
Surv(t2, d3 == 1) ~ group + da + dc + dp + z1 + z2 + z3 + z4 +
    z5 + z6 + z7 + z8 + z9


        Df AIC
- group  1 713
- z4     1 713
- z2     1 714
- da     1 714
- z5     1 714
- z1     1 714
- z3     1 714
- z7     1 714
- z6     1 714
<none>     715
```

```
- z8      1 717
- z9      1 720
- dc      1 733
- dp      1 734


Step:  AIC=713.3
Step:  AIC=702.8
Surv(t2, d3 == 1) ~ dc + dp + z1 + z8 + z9


        Df AIC
<none>     703
- z1     1 705
- z8     1 707
- z9     1 707
- dc     1 718
- dp     1 724
> leukemia_forward_AIC <- step(null,scope=list(lower=null,
+                          upper=full),data=bmt2,
+                          direction="forward")
Start:  AIC=746.6
Surv(t2, d3 == 1) ~ 1


         Df AIC
+ dp     1 724
+ dc     1 734
+ z8     1 741
+ z9     1 745
+ group  1 746
<none>     747
+ z2     1 747
+ z3     1 748
+ z1     1 748
+ z5     1 748
+ da     1 748
+ z7     1 749
+ z6     1 749
+ z4     1 749


Step:  AIC=724.2
Step:  AIC=702.8
Surv(t2, d3 == 1) ~ dp + dc + z8 + z9 + z1


         Df AIC
<none>     703
+ z7     1 704
```

```
+ da      1 704
+ z2      1 704
+ z6      1 704
+ z3      1 705
+ z5      1 705
+ group   1 705
+ z4      1 705
```

We look at the coefficients obtained the three methods:

```
> coefficients(leukemia_both_AIC)
      dc        dp        z1        z8        z9
-0.97295 -1.68763   0.02599   0.58696 -0.26861
> coefficients(leukemia_back_AIC)
      dc        dp        z1        z8        z9
-0.97295 -1.68763   0.02599   0.58696 -0.26861
> coefficients(leukemia_forward_AIC)
      dp        dc        z8        z9        z1
-1.68763 -0.97295   0.58696 -0.26861   0.02599
```

In each of the three techniques, we have ended with the same model. □

As with the parametric model selection and the method of computing the BIC, we can augment the **step** function with the option of **k = log()** which will then lead to the model selection based on the BIC. Note that the software will still continue to display the criterion as **AIC** only. However, this is a simple inconvenience that we will have to live with.

Example 33 *Model Selection with Cox Proportional Hazards Model Based on BIC.*

```
> # BIC Analysis
> leukemia_both_BIC <- step(leukemia_coxph,direction = "both",
+                           k = log(leukemia_coxph$nevent))
Start:  AIC=746.8
Surv(t2, d3 == 1) ~ group + da + dc + dp + z1 + z2 + z3 + z4 +
    z5 + z6 + z7 + z8 + z9


          Df AIC
- group    1 742
- z4       1 742
- z2       1 743
- da       1 743
- z5       1 743
- z1       1 743
- z3       1 743
- z7       1 743
```

```
- z6       1 744
- z8       1 746
<none>       747
- z9       1 749
- dc       1 762
- dp       1 763


Step:  AIC=742.4
Step:  AIC=714.9
Surv(t2, d3 == 1) ~ dc + dp + z1 + z8 + z9


        Df AIC
<none>       715
- z1       1 715
- z8       1 717
- z9       1 717
+ z7       1 718
+ da       1 719
+ z2       1 719
+ z6       1 719
+ z3       1 719
+ z5       1 719
+ group    1 719
+ z4       1 719
- dc       1 728
- dp       1 734
> leukemia_back_BIC <- step(full,data=bmt2,
+                        k = log(leukemia_coxph$nevent),
+                        direction = "backward")
Start:  AIC=746.8
Surv(t2, d3 == 1) ~ group + da + dc + dp + z1 + z2 + z3 + z4 +
    z5 + z6 + z7 + z8 + z9


          Df AIC
- group    1 742
- z4       1 742
- z2       1 743
- da       1 743
- z5       1 743
- z1       1 743
- z3       1 743
- z7       1 743
- z6       1 744
- z8       1 746
<none>       747
```

```
- z9      1 749
- dc      1 762
- dp      1 763


Step:   AIC=742.4
Step:   AIC=714.9
Surv(t2, d3 == 1) ~ dc + dp + z1 + z8 + z9


         Df AIC
<none>      715
- z1      1 715
- z8      1 717
- z9      1 717
- dc      1 728
- dp      1 734
> leukemia_forward_BIC <- step(null,scope=list(lower=null,
+     upper=full),data=bmt2,k = log(leukemia_coxph$nevent),
+     direction="forward")
Start:  AIC=746.6
Surv(t2, d3 == 1) ~ 1


         Df AIC
+ dp      1 727
+ dc      1 736
+ z8      1 743
<none>      747
+ z9      1 747
+ group   1 748
+ z2      1 750
+ z3      1 750
+ z1      1 750
+ z5      1 751
+ da      1 751
+ z7      1 751
+ z6      1 751
+ z4      1 751


Step:   AIC=726.6



Step:   AIC=714.8
Surv(t2, d3 == 1) ~ dp + dc + z8


         Df AIC
<none>      715
```

```
+ z9      1 715
+ z2      1 717
+ z1      1 717
+ z5      1 718
+ z7      1 718
+ da      1 719
+ group   1 719
+ z6      1 719
+ z4      1 719
+ z3      1 719
> coefficients(leukemia_both_BIC)
       dc        dp        z1        z8        z9
-0.97295 -1.68763   0.02599   0.58696 -0.26861
> coefficients(leukemia_back_BIC)
       dc        dp        z1        z8        z9
-0.97295 -1.68763   0.02599   0.58696 -0.26861
> coefficients(leukemia_forward_BIC)
      dp        dc        z8
-1.7221  -0.8627   0.4897
```

The bi-directional and backward selection procedures based on BIC return the same model as before. However, the forward selection method driven by BIC gives a model with only dp, dc, and z8 as the best model. Which model should be selected then? The answer is to look at their BIC values:

```
> AIC(leukemia_both_BIC)
[1] 702.8
> AIC(leukemia_back_BIC)
[1] 702.8
> AIC(leukemia_forward_BIC)
[1] 707.5
```

Hence, we pick the five variables.   □

An alternative to the AIC and BIC is emerging over the past decade and half, and it is the topic of discussion in the next section.

7.3 FIC Selection

When the AIC and BIC are applied on the Cox proportional hazards model, it works only on the partial likelihood function. Recall that the baseline hazard rate is not part of the partial likelihood. Claeskens and Hjort in a series of four papers from 2003 to 2008 proposed a new criterion called as focused information criterion. The critical point observed by them was that the information criteria are not concerned with the real purpose of the selected model. For instance, if the purpose of the study is toward prediction, the information criterion should take this into account.

What does focus mean here? The set of covariates identified by the AIC and BIC criteria is based on the (partial) likelihood function. When the parameter of interest is a different quantity, say median survival time, the covariates given by the two popular information criteria might no longer be the best. Moreover, we are not aware of invariant properties of the models identified by the criteria toward the functions of interest. The Focused Information Criterion, FIC, selects variables in the model depending on the problem on hand. For instance, the FIC selected variables with the objective of modeling for median survival time will be different from those where the purpose is analysis of survival curves depending on the treatments received by the patients. The method also takes model averaging into account.

The theory behind FIC is not within the scope of the chapter and we refer to the series of papers by Claeskens and Hjort that begin with the focus on linear regression models and then work their way with the Cox proportional hazards model to the Aalen's linear additive regression model. Claeskens and Hjort (2008)[26] monograph provides comprehensive details on the technique. We will settle with an application of the FIC criterion using `fic` function from the R package of same name.

We will illustrate use of FIC with the bone marrow transplant dataset available in `KMsurv` package.

Example 34 *Focused Information Criterion with Cox PH Model.*
We consider only six covariates and consider the problem of selecting the best model when the focus is on the survival function. The `fic` *function requires quiet a few data tweaks.*

The full model is first setup using the `coxph` *function. Next, all possible model combinations are generated by using the* `all_inds` *function on the indicator varaible of the covariate being selected. Keeping* `group` *as a base variable, we have* $2^5 = 32$ *possible models. This is easily seen with the R code block starting with* `inds0 <- expand_inds` *up to the check point* `dim(combs)`. *If we want* `group` *as a variable for selection, the number of possible models is* $2^6 = 64$, *seen in the code block from* `inds_all` *to* `dim(combs_all)`.

The covariate matrix needs to be explicitly fed to the `fic` *function. Thus, we use the function* `newdata_to_X` *and then run the* `fic` *function with* `"survival"` *as the focus for model selection.*

```
> data(bmt)
> bmt2 <- bmt[,c("t2","d3","group","da","dc","dp","z1","z2")]
> bmt_ph2 <- coxph(Surv(t2,d3==1)~.,data=bmt2)
> inds0 <- expand_inds(c(1,0,0,0,0,0),bmt_ph2)
> inds0
     [,1] [,2] [,3] [,4] [,5] [,6]
```

```
[1,]    1    0    0    0    0    0
> combs <- all_inds(bmt_ph2,inds0)
> combs
       group da dc dp z1 z2
100000     1  0  0  0  0  0
110000     1  1  0  0  0  0
101000     1  0  1  0  0  0
111000     1  1  1  0  0  0

100111     1  0  0  1  1  1
110111     1  1  0  1  1  1
101111     1  0  1  1  1  1
111111     1  1  1  1  1  1
> dim(combs)
[1] 32  6
> inds_all <- expand_inds(c(0,0,0,0,0,0),bmt_ph2)
> combs_all <- all_inds(bmt_ph2,inds_all)
> dim(combs_all)
[1] 64  6
> bmt_X2 <- subset(bmt3,select=-c(t2,d3))
> bmt_new_X2 <- newdata_to_X(bmt_X,bmt_ph2,intercept = FALSE)
> bmt_ficall2 <- fic(bmt_ph2,inds=combs,inds0=inds0,
+                    focus = "survival",X=bmt_new_X2,t=500)
> save(bmt_ficall2,file="../Output/bmt_ficall2.RData")
> pdf("../Output/FIC2_BMT_Survival.pdf",height=15,width=15)
> ggplot_fic(bmt_ficall2)
> dev.off()
null device
          1
> min(bmt_ficall2$FIC)
[1] 5.23e-05
> which.min(bmt_ficall2$FIC)
[1] 4295
> bmt_ficall2$mods[which.min(bmt_ficall2$FIC)]
[1] 101100
32 Levels: 100000 110000 101000 111000 100100 ... 111111
> modname <- as.character(
+     bmt_ficall2$mods[which.min(bmt_ficall2$FIC)])
> combs[which(rownames(combs)==modname),]
       group da dc dp z1 z2
101100     1  0  1  1  0  0
```

The best model selected by the FIC is the one with covariates group, dc, *and* dp.   □

The FIC has been extended to the additive hazards model in Hjort (2008)[53].

7.4 Penalization with the Proportional Hazards Model

Likelihood-based criteria are not the only way to select a model. Penalization techniques are an useful alternative. This method is useful in not only restricting the undue importance of the covariates, but also, in certain forms, it helps averting the impact of multicollinearity. The standard requirement, is that, we need $\mathbf{X}$ to be of full rank. When the full rank condition is not satisfied, it implies that one of the columns of the regression matrix is a linear combination of other covariates. Consequently, the determinant of $\mathbf{X'X}$ becomes zero. In real world, with data, the determinant becomes nearly equal to zero, and not exactly zero, implying that the inversion will take place. Consequently, the estimated coefficients will be large and unstable. Before embarking on the ways of overcoming the problem through penalization, we will try to fit the Cox proportional hazards regression model to the case where the number of covariates is nearly half the number of available observations.

Example 35 *Netherlands Cancer Institute 70 Genes. The dataset has been introduced earlier in Section 1.2.6. Here, metastasis-free follow-up time is noted for the lymph node positive breast cancer patients. Information on five explanatory variables—patient's age, grade of the tumor, estrogen receptor status, diameter of the tumor, and number of affected lymph nodes,—are captured. Besides these, measurements for 70 prognostic gene variables are also recorded. Further, only 48 patients experience the event of interest: death or relapse of metastasis.*

We will model for the time to metastasis/death by all available covariates and then look at their estimated coefficients.

```
> data("nki70")
> dim(nki70)
[1] 144  77
> summary(nki70$time)
    Min.  1st Qu.  Median    Mean  3rd Qu.     Max.
 0.05476  4.70568  6.99521 7.35130  9.98631 17.65914
> sum(nki70$event)
[1] 48
> nk_simple_ph <- coxph(Surv(time,event)~.,nki70)
Warning message:
```

```
In fitter(X, Y, istrat, offset, init, control, weights = weights,  :
  Ran out of iterations and did not converge
> sapply(nki70[,8:77],range)
       TSPYL5 Contig63649_RC  DIAPH3  NUSAP1 AA555029_RC
[1,] -1.0828        -0.5077 -0.6789 -0.7968     -0.4778
[2,]  0.6018         0.7757  0.6178  0.5067      0.3741


      EGLN1 NM_004702    ESM1 C20orf46
[1,] -0.5036   -0.6677 -0.8972  -0.4506
[2,]  0.4062    0.7101  0.8474   0.9915
> sort(coefficients(nk_simple_ph),decreasing = TRUE)
       C9orf30      IGFBP5.1            AP2B1
       444.733       280.808          206.813
        DIAPH3        COL4A2             PECI
       205.904       156.766          134.282


        PITRM1          BBC3         AYTL2        SERF1A
      -313.663      -340.703      -342.653      -370.865
```

The consequence of fitting the Cox proportional hazards model for all co-variates leads to a warning message. The numerical algorithm did not achieve convergence by the time the iterations went dry. When we look at the range of the covariate values, most range in the interval $[-1, 1]$. Also, the time to follow up for the patients varies from nearly 0 to a maximum of 18. However, the coefficients are having much larger value, three digit numbers. This is a possible indication that the fit is pretty poor and the covariates are having ununsual impact.

The curious reader should experiment by increasing the number of iterations with the optio of `iter.max`, say `coxph(Surv(time,event)~.,nki70)`. What is going wrong over there?

Now, we will rebuild the model by dropping the five non-gene measurement covariates, and redo the analysis.

```
> nk_simple_ph2 <- coxph(Surv(time,event)~.,nki70[,c(1,2,8:77)])
> sort(coefficients(nk_simple_ph2),decreasing = TRUE)
          PECI          PRC1    PALM2.AKAP2
      24.98770      15.66371       10.76001
          MELK   RP5.860F19.3      NM_004702
       9.93109       8.79607        7.90260


      DIAPH3.2        STK32B         SLC2A3        PECI.1
     -19.66752     -20.96568      -22.35626     -22.69578
```

The numerical approximation method has achieved convergence, which is an achievement with the experimentation over the choice of covariates. Are the *coefficient values acceptable? The reader should answer the question for his own inquisitive understanding.* □

How does penalization help fix the 'singular' problem? Singularity problem means that the determinant of the Gram matrix of $\mathbf{X}$ is nearly zero. The method of penalization overcomes zero determinant problem. It is easier to understand it in the context of linear regression model. With the generic linear regression, we have $Y = g(\boldsymbol{\beta}'\mathbf{x}) = \boldsymbol{\beta}'\mathbf{x}$, the unit link function being in action $g(u) = u$, the estimate of the regression coefficient is $\hat{\boldsymbol{\beta}} = (\mathbf{X}'\mathbf{X})^{-1}\mathbf{X}'\mathbf{Y}$. The ordinary-least squares estimate is an optimal solution of the following problem:

$$\operatorname*{argmin}_{\boldsymbol{\beta}} \left\{ \boldsymbol{\epsilon}'\boldsymbol{\epsilon} = \left(\mathbf{Y} - \boldsymbol{\beta}'\mathbf{X}\right)' \left(\mathbf{Y} - \boldsymbol{\beta}'\mathbf{X}\right) \right\}.$$

The coefficient vector $\boldsymbol{\beta}$ is obtained as a solution to the unconstrained optimization problem in the sense that we have not capped the decision vector $\boldsymbol{\beta}$. Consequently, there is no limit on the β's and we might end up with high values as seen in Example 35. In the ℓ_2 penalty, we instead minimize by the following penalization criteri:

$$\operatorname*{argmin}_{\boldsymbol{\beta}} \left\{ \boldsymbol{\epsilon}'\boldsymbol{\epsilon} - \lambda \sum_{j=1}^{p} \beta_j^2 \right\}. \tag{7.3}$$

Note that the intercept term β_0 is kept out of minimization problem. Here, λ is the Lagrangian multiplier. For λ values closer to zero, there is no penalization and the solution would be nearly equivalent to the usual OLS solution. For extremely large values of λ, the only way minimization is achieved is through values of β's closer to zero. Thus, we can intuitively see that for nonnegative values of λ, we are constraining the search over the size of regression coefficients β's. Also, the optimization problem in Equation 7.3 is equivalent to the following constrained search:

$$\operatorname*{argmin}_{\boldsymbol{\beta}} \{\boldsymbol{\epsilon}'\boldsymbol{\epsilon}\}$$
$$\texttt{subject to}$$
$$\|\boldsymbol{\beta}\|_2 \le s.$$

The ℓ_2 penalty is imposed by the norm $\|.\|_2$ which for a given vector $\mathbf{z} = (z_1, \ldots, z_p)$ is defined by $\|\mathbf{z}\|_2 = \sum_{j=1}^{p} z_j^2$.
    A solution of the above ℓ_2 penalty is

$$\hat{\boldsymbol{\beta}} = (\mathbf{X}'\mathbf{X} - \lambda\mathbf{I})^{-1}\mathbf{X}'\mathbf{Y}.$$

The form of $(\mathbf{X}'\mathbf{X} - \lambda\mathbf{I})^{-1}$ might be familiar to the reader in multivariate analysis and the solution utilizes λ as the eigen values and their associated eigen vectors. Consequently, the solution leads to the use of *principal components*, and the analysis is known as *principal component analysis*, PCA.

The principal components are linear combinations of the covariates. It is well known that the principal components are uncorrelated and hence the use of the ℓ_2 penalty replaces the regular inverse $(\mathbf{X}'\mathbf{X})^{-1}$ by the principal components. Since the principal components are uncorrelated, the fundamental assumption of the linear independence of the covariates is automatically met, see Seber and Wild (1989). In the context of linear regression analysis, the OLS and ML estimates coincide. However, we do not use the least squares solution when it comes to the relative risk model.

Recollect the form of the intensity process of the relative risk model: $h(t|\mathbf{x}(t)) = Y(t)h_0(t)r(\boldsymbol{\beta}'\mathbf{x}(t))$. Based on n observations, the partial likelihood is given by, see Equation 5.11,

$$L(\boldsymbol{\beta}|\texttt{data}) = \prod_{T_i} \frac{r(\boldsymbol{\beta}, \mathbf{x}_{m_i}(T_i))}{\sum_{l \in \mathcal{R}_i} r(\boldsymbol{\beta}, \mathbf{x}_l(T_i))}.$$

The ℓ_2 penalty for the relative risk model is now based on determining $\boldsymbol{\beta}$ which will maximize the partial likelihood function, as against minimizing $\boldsymbol{\epsilon}'\boldsymbol{\epsilon}$ earlier, with the Lagrangian multiplier λ:

$$\underset{\boldsymbol{\beta}}{\operatorname{argmax}} \left\{ L(\boldsymbol{\beta}|\texttt{data}) - \lambda \left\| \boldsymbol{\beta} \right\|_2 \right\}, \tag{7.4}$$

where $\|\boldsymbol{\beta}\|_2 = \sum_{j=1}^p \beta_j^2$. An important problem in determining the regression coefficient vector $\boldsymbol{\beta}$ under the penalization methods, ℓ_2 now and ℓ_1 later, is the choice of the Lagrangian multiplier. In the case of a linear regression model, determination of determining the Lagrangian multiplier can be done using any one of the following methods: (i) Hoerl, Kennard, and Baldwin, (ii) McDonald and Galarneau, and (iii) Mallows. The reader might refer to Chapter 9 of Montgomery, et al. (2021)[84] for the discussion and implementation.

It is not mandatory for the practitioner to penalize all the coefficients. Penalization of sub-vectors is allowed. However, we skip over this feature of penalization techniques.

In the context of the relative risk model, and other nonlinear methods, we resort to a grid search method of various λ values and select the one which minimizes the AIC, or other related quantities. Using the `penalized` package, we illustrate it for the Netherlands Cancer Insititute data set.

Example 36 *The L^2-penalty for Netherlands Cancer Institute Data.*
The `penalized` *function can be applied over a* `Surv` *object. We need to specify which covariates need to be penalized under the option of* `penalized` *within that function, and the variables that need to be left out as unconstrained under the choice of* `unpenalized`. *To carry out the ℓ_2 penalty, we use the option of* `lambda2`. *Note that we do not have ML estimate of λ. We will briefly explain after the example the use of grid search to determine λ. For now, we will use $\lambda = 5$ and specify in the program as* `lambda2=5`. *Why is the option named as* `lambda2` *and not* `lambda` *or* `lambda1`? *We will get the answer in what follows.*

We will begin with looking at how many coefficients estimated in Example 35 exceed 2 in magnitude, and then fit the relative risk model with the claimed penalty factor.

```
> sum(abs(coefficients(nk_simple_ph2)>2))
[1] 24
> # The L-2 Penalty
> nk_l2_ph <- penalized(Surv(time,event),penalized=nki70[,8:77],
+                       data = nki70,lambda2=5
+                       )
> nk_l2_ph
Penalized cox regression object
70 regression coefficients

Loglikelihood =  -232.9
L2 penalty =  10.6  at lambda2 =  5
> sum(abs(coefficients(nk_l2_ph)>2))
[1] 0
> sort(round(coefficients(nk_l2_ph),4))
          ZNF533            MS4A7           GPR180
         -0.5840          -0.3931          -0.3681
          SCUBE2            RAB6B          RTN4RL1
         -0.3323          -0.3032          -0.2933

         QSCN6L1         IGFBP5.1           IGFBP5
          0.3966           0.4674           0.4887
            PRC1
          0.5811
```

In the earlier instance, we had 24 out of 70 genes which had an estimated coefficient value greater than 2. With the choice of the penalty factor of $\lambda = 5$, we penalized the relative risk model and obtained the coefficients under the constrained environment. Now, we do not have any coefficient value greater than 2 which shows that the influence of the variables has been shrunk.

$\square$

How to choose λ? To obtain the correct value of λ in certain sense, note that the log-likelihood value is stored in the penalized R object, and in the previous case it is `Loglikelihood = -232.9`. The AIC/BIC criterion can then be easily obtained for the fitted model. Varying λ values over a grid, the search technique can be easily looped in and the best fit model be obtained.

Tibshirani (1996)[118] gave an important breakthrough with his invention that unifies penalization as well as model selection. Tibshirani invented the method of *Least Absolute Shrinkage and Selection Operator* and it is more popularly known these days by its abbreviation to the extent that it has also become a word and often called as "LASSO". The penalty used here is the ℓ_1-penalty, or the $\|.\|_1$ norm which for a given vector $\mathbf{z}$ is defined by $\|\mathbf{z}\|_1 = \sum_{j=1}^{p} |z_j|$. Here, the shrinkage happens using the absolute values. The reason for the key terms of 'selection operator' is for a good statistical reason. The ℓ_1 penalty forces some of the β_j's to be exactly equal to zero. This phenomenon does not occur in case of the ℓ_2 penalty. In our previous example, a simple check of sum(coefficients(nk_12_ph)==0) returns the output as 0 which means, ignoring statistical significance, that the covariates have some impact on the hazard rate. If the β_j's are coerced to be zero, it is as good as eliminating the variable from the model and hence the moniker of 'selection operator' holds perfect for LASSO. The likelihood function is now maximized with the ℓ_1 penalty:

$$\underset{\boldsymbol{\beta}}{\operatorname{argmax}} \left\{ L(\boldsymbol{\beta}|\texttt{data}) - \lambda \|\boldsymbol{\beta}\|_1 \right\}. \tag{7.5}$$

The LASSO technique requires immense computational power, and we abstain from the related discussion. It is important to note that LASSO is a much versatile procedure and in many machine learning, discussed elaborately in the next part of the book, it is the default and efficient way of selecting variables. It is common to have humongous number of variables in the 'Big Data' paradigm and often LASSO is the only variable selection procedure option.

The LASSO adaptation for the relative risk model has been done by Goeman (2010)[47] and he is also the creator and maintainer of the penalized package. The option of lambda1 in the penalized function is for the ℓ_1 penalty, or equivalently the LASSO. Let us continue to apply the penalized function.

Example 37 *The LASSO Penalty for Netherlands Cancer Institute Data with Cox PH Model. The next line of R code does the required computation and fits the LASSO penalty for the relative risk model. We try with the λ value of 2:*

```
> # The LASSO Penalty
> nk_l1_ph <- penalized(Surv(time,event),penalized=nki70[,8:77],
+                       data = nki70,lambda1=2
+                       )
# nonzero coefficients: 16
```

The console output says that we are left with 16 nonzero coefficients which is the same as saying that $70 - 16 = 54$ gene variables have no impact on the time to metastasis or death. This reduction is much wanted relief.

```
> nk_l1_ph
Penalized cox regression object
70 regression coefficients of which 16 are non-zero

Loglikelihood =   -239.2
L1 penalty =   14.35  at lambda1 =  2
> sort(round(coefficients(nk_l1_ph),4))
```

ZNF533	GPR180	MS4A7
-0.6827	-0.4385	-0.2828
KNTC2	RAB6B	Contig20217_RC
-0.2324	-0.1774	-0.1559
SCUBE2	HRASLS	IGFBP5.1
-0.1269	-0.0761	0.1265
MMP9	Contig63649_RC	ESM1
0.1751	0.2473	0.5635
QSCN6L1	Contig32125_RC	IGFBP5
0.5770	0.6361	0.7892
PRC1		
1.8884		

The extracted loglikelihood value is useful in the grid search for the value of λ that maximizes the likelihood function. □

A final closure remark is in order before we close the discussion on penalty methods for the relative risk model. The choice of the norm is as much a source of confusion as the specific value of the penalty factor. Especially, we do not have a practical interpretation of the penalties and therefore it becomes difficult to explain the choice of the norm. Practitioners are also split because of this reason. While the LASSO force fits some values of the coefficients to be exactly zero, the ℓ_2 penalty looks more balanced by being symmetric in nature. Thus, it is even better to combine both the norms instead of getting tied down in selecting one of them. The ℓ_1 and ℓ_2 penalties can be combined to solve the following problem:

$$\underset{\boldsymbol{\beta}}{\text{argmax}} \left\{ L(\boldsymbol{\beta}|\texttt{data}) - \lambda_1 \left\| \boldsymbol{\beta} \right\|_1 - \lambda_2 \left\| \boldsymbol{\beta} \right\|_2 \right\}. \tag{7.6}$$

The combination of both the penalties is known as *elastic net* and it benefits from the selection strategy of ℓ_1 as well as the smooth penalty of ℓ_2.

7.5 Penalization with Aalen's Semiparametric Hazards Model

The Aalen's additive hazards model is an important alternative to the relative risk model, especially when the proportional hazards assumption is violated. The intensity process form of the Aalen's additive model is $h_i(t) = Y_i(t) \left\{ \beta_0(t) + \beta_1(t)x_{i1}(t) + \ldots + \beta_p(t)x_{ip}(t) \right\}, i = 1, \ldots, n$. We saw in Section 6.2 that the cumulative regression function is estimated by

$$\hat{\mathbf{B}}(t) = \int_0^t J(s) \mathbf{X}^-(s) d\mathbf{N}(s) = \sum_{T_i \leq t} J(T_i) \mathbf{X}^-(T_i) \Delta \mathbf{N}(T_i).$$

Though statistical inference is possible in the context of the additive regression model, model selection remains a problem and we are not aware of an algorithm and technique which accomplishes this task. For instance, the `step` function works for the `coxph` fitted models whereas the same function does not apply or work for the fitted Aalen additive regression model, the class `aalen`. Contrast the regression coefficient function $\beta(t)$ against the simple time-free unknown vector in the relative risk model β. The model corresponding to the former has been referred to in the text as fully nonparametric model while the latter is referred to as semiparametric model. Recollect the form of the semiparametric model associated with the additive hazards model given in Equation 6.8 as $h_i(t) = Y_i(t)\{\beta_0(t) + \beta'(t)\mathbf{x}_i(t) + \gamma'\mathbf{z}_i(t)\}$. We have fitted a semiparametric model for the PBC dataset in Example 27. We would like the reader to recollect that the regression coefficients associated with $\mathbf{z}_i(t)$ in that example were nearly equal to zero, that is, $\hat{\gamma} \approx 0$.

Lin and Yin (1994)[72] proposed a semiparametric model of the additive hazards regression model whose intensity process is of the following form:

$$h_i(t) = Y_i(t)\{\beta_0(t) + \gamma'\mathbf{z}_i(t)\}. \tag{7.7}$$

This is the semiparametric additive model that can be compared with the relative risk model in terms of the coefficient values. The coefficients γ cannot be obtained by using the OLS method. Before embarking on further discussion of penalization and model selection with the semiparametric additive model, we will first fit it to the PBC dataset.

The Lin and Ying's semiparametric model is obtained using the `ahaz` R package. As on date, ties are not supported in the current implementation and they need to be separated explicitly. We accomplish that using the `jitter` function which adds small noise to the actual times. We do not lose a lot by using this unconventional trick.

Example 38 *Fitting Semiparametric Additive Hazards Model. We continue with the PBC dataset analysis. In the* `ahaz` *implementation, we need to specify the* `Surv` *object and the covariate matrix separately. Observations with missing covariate values are not omitted by the R function ahaz and we need to remove them by using* `na.omit` *function. Rest of the program is straightforward.*

```
> # Fitting Semiparametric Additive Hazards Regression Model
> data(pbc)
> pbc2 <- subset(pbc,select=c(time,status,trt,age,sex,
+                             edema,bili,albumin,protime))
> pbc2 <- na.omit(pbc2)
> X <- as.matrix(subset(pbc2,select=c(age,edema,bili,
+                             albumin,protime)))
> set.seed(123)
> pbc_time <- jitter(pbc2$time)
```

```
> pbc_surv <- Surv(pbc_time,pbc2$status==2)
> pbc_ahaz <- ahaz(pbc_surv,X)
> summary(pbc_ahaz)

Call:
ahaz(surv = pbc_surv, X = X)

  n = 312


Coefficients:
          Estimate Std. Error Z value Pr(>|z|)
age        6.31e-06   1.83e-06    3.45  0.00057 ***
edema      5.14e-04   1.96e-04    2.62  0.00872 **
bili       6.48e-05   1.35e-05    4.80  1.6e-06 ***
albumin   -2.24e-04   6.49e-05   -3.45  0.00057 ***
protime    5.76e-05   2.23e-05    2.58  0.00975 **
---
Signif. codes:
0 '***' 0.001 '**' 0.01 '*' 0.05 '.' 0.1 ' ' 1

Wald test = 57  on 5 df,    p=5.05e-11
```

*Note that the p-value 5.05e-11 refers to the overall model. Here, all the covariates have turned out to be significant. Note that if we round off the coefficient values at four digits, except for **edema**, all of them would be returned as zero. The implication is not correct though. However, the interpretation is related to the hazard rate, and not the time to event of interest. In that way, it makes sense for the coefficients to be much smaller.* □

Penalization methods are not generally available for regression functions, such as $\beta(t)$s. Now that we have the coefficients for the additive regression model, we look at the LASSO penalty. The **ahaz** package has a function **ahazpen** which implements the LASSO penalty for the semiparametric model 7.7. For more theory and details, the reader might refer to Martinussen and Scheike (2006)[76] and Gorst-Rasmussen and Scheike (2012)[49].

Example 39 *The LASSO Penalty for Netherlands Cancer Institute Data with Aalen's Semiparametric Regression Model. We will fit Aalen's semiparametric additive hazards model for the metastasis event discussed earlier in the chapter. The **penalty** option offers multiple choices of penalty in the **ahazpen** function. Here, we do not have to specify a choice for λ of the ℓ_1 penalty. The reader may try to find the choice of the penalty factor.*

```
> # The LASSO Penalty for the Netherlands Cancer Institute
> data("nki70")
```

```
> set.seed(123)
> nktime <- jitter(nki70$time)
> nk_surv <- Surv(nktime,nki70$event)
> nkX <- as.matrix(nki70[,8:77])
> nk_lasso <- ahazpen(nk_surv,nkX,penalty = "lasso")
> nk_lasso

Call:
ahazpen(surv = nk_surv, X = nkX, penalty = "lasso")

* No. predictors:                       70
* No. observations:                     144
* Max no. predictors in path:           70
* Penalty parameter lambda:
    -No. grid points:      100
    -Min value:          2.08e-05
    -Max value:            0.208
> round(nk_lasso$lambda,5)
  [1] 0.20773 0.18928 0.17247 0.15714 0.14318 0.13046 0.11887
  [8] 0.10831 0.09869 0.08992 0.08193 0.07466 0.06802 0.06198

 [92] 0.00004 0.00004 0.00004 0.00003 0.00003 0.00003 0.00003
 [99] 0.00002 0.00002
> dim(nk_lasso$beta)
[1]   70 100
```

The output shows that we have 70 predictors/covariates and 144 observations.
The hundred different values of λ is the grid search from 0.20773 to 0.00002.
At five decimals rounding off, we can find which penalty values are used in this
run. Next, for each of these penalties, we have the beta coefficients as seen by
running dim(nk_lasso$beta).

```
> nk_lasso$lambda[1];nk_lasso$beta[,1]
[1] 0.2077
 [1] 0 0 0 0 0 0 0 0 0 0 0 0 0 0 0 0 0 0 0 0 0 0 0 0 0 0 0
[28] 0 0 0 0 0 0 0 0 0 0 0 0 0 0 0 0 0 0 0 0 0 0 0 0 0 0 0
[55] 0 0 0 0 0 0 0 0 0 0 0 0 0 0 0 0
> nk_lasso$lambda[10];nk_lasso$beta[,10]
[1] 0.08992
 [1]   0.00000 0.00000 0.00000 0.00000 0.00000 0.00000
 [7]   0.02196 0.00000 0.00000 0.00000 0.00000 0.00000

[55]   0.00000 0.00000 0.00000 0.00000 0.00000 0.02169
[61]   0.00000 0.00000 0.00000 0.05403 0.00000 0.00000
[67]   0.00000 0.00000 0.00000 0.00000
> nk_lasso$lambda[20];nk_lasso$beta[,20]
```

```
[1] 0.03547
  [1]  0.000000  0.067079  0.000000  0.000000  0.000000
  [6]  0.000000  0.060324  0.000000  0.000000  0.082878


[61] -0.008092  0.000000  0.000000  0.142285 -0.052945
[66]  0.000000 -0.050575  0.016351  0.013289 -0.011621
> nk_lasso$lambda[50];nk_lasso$beta[,50]
[1] 0.002176
  [1]  0.005389  0.188746  0.028465 -0.018996  0.042190
  [6]  0.078264  0.097299 -0.023070 -0.010569  0.167670


[61] -0.029402  0.046618  0.021731  0.270086 -0.132067
[66]  0.034641 -0.052314  0.132945  0.018488 -0.028627
```

The above output gives us the coefficients for the corresponding values of λ. When the penalty is higher, every single coefficient is zero. As the penalty decreases, the number of nonzero coefficients increases. We can easily visualize the changes using the plot function.

```
> pdf("../Output/Netherlands_LASSO.pdf",height=5,width=8)
> plot(nk_lasso,xvar="lambda")
> dev.off()
RStudioGD
        2
```

Output of the above code is Figure 7.1. We can clearly see verification of the earlier claim that as the value of the penalty factor increases, the number of variables with estimated regression coefficients close to zero also increases. Using an appropriate cutoff, we can choose the penalty value and then their corresponding regression coefficients βs. □

An alternative to the LASSO penalty is provided by *Smoothly Clipped Absolute Deviation* penalty, or SCAD penalty introduced by Fan and Li (2001)[43] and adapted for the Aalen's semiparametric model in Gorst-Rasmussen's R package **ahaz**. We will simply illustrate it for the metastatis data.

Example 40 *The SCAD Penalty for Netherlands Cancer Institute Data with Aalen's Semiparametric Regression Model. The working of the SCAD is same as LASSO and we simply need to switch to the choice of* sscad *instead of* lasso *in the* penalty *option.*

```
> # The SCAD Penalty for the Netherlands Cancer Institute
> nk_scad <- ahazpen(nk_surv,nkX,penalty = "sscad")
> pdf("../Output/Netherlands_SCAD.pdf",height=5,width=8)
> plot(nk_scad,xvar="lambda")
> dev.off()
```

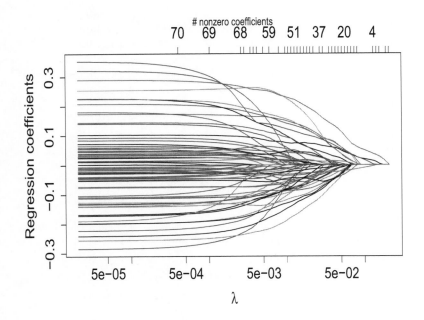

FIGURE 7.1: LASSO Penalization with Aalen's Semiparametric Regression Model

```
RStudioGD
        2
> nk_scad$lambda[1];nk_scad$beta[,1]
[1] 0.3005
 [1] 0 0 0 0 0 0 0 0 0 0 0 0 0 0 0 0 0 0 0 0 0 0 0 0 0 0 0
[28] 0 0 0 0 0 0 0 0 0 0 0 0 0 0 0 0 0 0 0 0 0 0 0 0 0 0 0
[55] 0 0 0 0 0 0 0 0 0 0 0 0 0 0 0 0
> nk_scad$lambda[10];nk_scad$beta[,10]
[1] 0.1301
 [1]   0.000000   0.023811   0.000000   0.000000   0.000000
 [6]   0.000000   0.000000   0.000000   0.000000   0.010270

[61]   0.000000   0.000000   0.000000   0.167665   0.000000
[66]   0.000000   0.000000   0.000000   0.000000   0.000000
> nk_scad$lambda[20];nk_scad$beta[,20]
[1] 0.0513
 [1]   0.000000   0.126066   0.000000   0.000000   0.000000
 [6]   0.000000   0.033530   0.000000   0.000000   0.160046

[61]  -0.021508   0.000000   0.000000   0.255495  -0.105238
```

```
[66]   0.000000   0.000000   0.225232   0.000000  -0.003199
> nk_scad$lambda[50];nk_scad$beta[,50]
[1] 0.003148
 [1]   0.012976   0.209121   0.051046  -0.043494   0.045689
 [6]   0.091272   0.103767  -0.040373  -0.014177   0.173625

[61]  -0.027828   0.066651   0.029720   0.286969  -0.126248
[66]   0.041629  -0.050104   0.144100   0.014638  -0.027287
```

Results of the above output are displayed in Figure 7.2. The SCAD and LASSO technique give us similar results. □

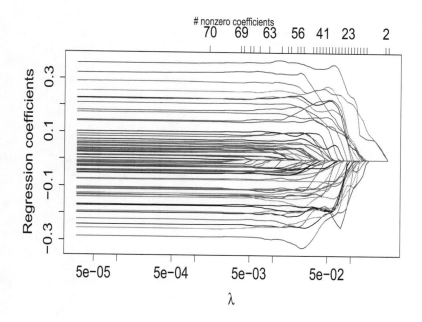

FIGURE 7.2: SCAD Penalization with Aalen's Semiparametric Regression Model

Penalization in the Aalen's semiparametric model is useful because penalization of the regression functions is an open problem.

7.6   Exercises

Exercise 7.1 *For each choice of forward, backward, and stepwise selection, and further for each of AIC and BIC, obtain the "best" model for the veterans's lung cancer study, see Exercise 4.3 for further details.* □

Exercise 7.2 *Find the best fit model for the PBC study under FIC.* □

Exercise 7.3 *Fit the ℓ_1 and ℓ_2 penalized models for the PBC data and compare them with the corresponding model fitted without any penalty.* □

Exercise 7.4 *The specification of penalization terms* `lambda1` *and* `lambda2` *in the* `penalized` *function might be further specified at the covariate level. Experiment with evaluation of the methods of penalty at covariate level and create a function to select the best model.* □

7.7 Más Lejos Temas

Model selection is a central idea in regression modeling. The possibility of finding the universally best model is incredibly difficult, and the role of *p*-values in the context of eliminating or keeping the variables in the model is subjective in the sense that the choice of the cut-off value for the *p*-value itself is not very clear. The AIC and BIC metrics are likelihood driven and they have been seen to keep the variables that might have seen the chopping block if driven by *p*-value. We always keep it in mind that the three different approaches of forward, backward, and stepwise selection may lead to different models. It will be seen in the later part that one might as well keep the top models, ranking by AIC values, and combine them in a meaningful way.

The FIC is a welcome addition because it is based on entire model and on the partial likelihood. It might be argued that the baseline/intercept is seldom a bothersome variable in model selection. However, the intercept term is barely the equivalent of the baseline hazard function which is heavily linked with time with the exception of exponential regression model. The better part of the FIC is that it can be driven from the perspective of prediction problem too discussion which is beyond the scope of the chapter. The reader may refer to Claeskens and Hjort (2008)[26] for the general details related to FIC. The technique appears to have debuted in Claeskens and Hjort (2003)[25], application to the Cox model was done in Hjort and Claeskens (2006)[54], and applied to the Aalen's model in Hjort (2008)[53].

Penalization methods are powerful alternatives to the conventional variable selection methods and we have seen them in action in Sections 7.4 and 7.5. The ℓ_2 and ℓ_1 penalties provide the practitioner with flexible options. The combination of the two penalties becomes the elastic net which has been mentioned and not illustrated. In fact, the R package `fastcox` implements the elastic net for the Cox proportional hazards model. The associated penalty factors of λ is often determined by a grid search over a cross-validation sample. The ℓ_1 and ℓ_2 penalties are sufficiently versatile in enhancing other instances of regression models. In fact, the usage of the penalties is more common in the *machine learning* domain.

It would be interesting to examine application of penalization to linear survival regression methods, or even in the context of the pseudo-observations modeling. To an extent, the AIC method might be extended by looking at equivalent likelihood representations. The purpose of model selection, or dimensionality reduction, often results in reduction in variance. Thus, the reader might experiment with the Box-Cox transformation, see Box and Cox (1964)[15].

Part II

Machine Learning Methods

Why Machine Learning?

Why Machine Learning? That question might have occurred in the reader's mind before embarking on a journey of the vast body of machine learning techniques. A number of questions arise naturally. Should one enlist the techniques of this new paradigm? Does it generally provide more accurate results over the conventional statistical methods? Should the authors begin with statistical methods first and then write on Machine learning? In fact, we do not have to try too hard to justify the inclusion of Machine Learning (ML) methods in the book. Before we give further justification of the need of the ML methods, a bit of diversion is in order.

Leo Breiman had completed his PhD under the supervision of the eminent probabilist Michel Loeve at Berkeley and his 1968 text simply titled 'Probability' became an all time classic. However, he turned his focus from the right arm (theory) of the subject to the left arm (applications) and then pioneered the invention of 'Classification and Regression Trees' (CART) in collaboration with other famous statisticians as Friedman, Olshen, and Stone. CART is an important spoke in the ML wheel. Sir D.R. Cox is one of the pillars of modern statistics and we have seen in Part I his major contributions to Survival Analysis. These two magnificient Statisticians had a great discussion in the paper *Statistical Modeling: The Two Cultures* see Breiman (2001)[18]. While Breiman bats for the ML paradigm, Sir Cox holds the ground for the century plus old subject. There is enough ground in common and one can complement the findings from one approach by using validation of others.

Bootstrap technique will play a crucial role in ML topics too. We would quickly see in this part how vital is the resampling procedure. Invented by Efron in 1979, the method has been marvelously modified by Brieman in extending CART to bagging and random forests. On a similar vein, the ℓ_1 penalty, aka LASSO, is important in ML where the notion of sampling distributions is subdued. The use of Statistics in ML, and vice versa, is so much interwined that Bickel and Doksum's 1977 classic revised in 2015 now has strong ML underpinnings. Similarly, Efron and Hastie (2016)[41] provides lot of statistical explanation behind the ML techniques. Besides, ML has a lot of rigorous theoretical underpinnings and it is beyond the scope of the current text to go into them.

The three main theories which form the backbone of ML are (i) No free-lunch theorem, (ii) Empirical Risk Minimization, and (iii) Vapnik-Chervonenkis dimension. Shalev-Shwartz and Ben-David (2014)[104] go into the details of the theoretical foundation of ML. The no free-lunch theorem asserts that there does not exist a single best optimization algorithm. As in statistics, optimization is required in ML too. By extension, the no free-lunch theorem says that there is no single best machine learning algorithm for all prediction scenarios.

Traditionally, in statistics, we almost always undertake an optimization problem, say minimizing the least squares, or maximizing the likelihood function. Under the assumption of a true, but unknown data generating mechanism, we try to approximate the unknown probability distribution using parametric, nonparametric, Bayesian, and other approaches. The error minimization strategy is with respect to the true probability distribution. In Machine Learning, the central principle is *empirical risk minimization*. The core idea is that since it is impossible to know the so-called true distribution, there is no point tracing it down. Instead, the risk/loss is best defined over observed data which is generally marked as *training data*. Thus, we are attempting to minimize the empirical risk.

The Vapnik-Chervonenkis dimension is a measure of the complexity, expressive power, richness, or flexibility, of a set of functions that can be learned by a statistical binary classification algorithm. We will close the monologue with an interesting observation of Vladimir Vapnik.

In his 1998 classic book, Vapnik says the following: "If you possess a restricted amount of information for solving some problem, try to solve the problem directly and never solve a more general problem as an intermediate step". What does this mean? Consider the classical linear discriminant analysis (LDA). Fisher's method is useful in classifying observations into one of two populations. Basically, given sufficient number of observations from the two populations, we create a mechanism which will obtain a score for a given observation based on a scaling of the pooled variance-covariance matrix and the distance of the observation from the center of the mean vector of one of the populations. An underlying assumption here is that the two populations follow multivariate normal distribution with a common variance-covariance matrix and different mean vectors. Now, according to Vapnik, when trying to solve the problem of classifying the observations to belong to one of the two populations, classical statistics, as well as the Bayesian methods, assumes the complex form of multivariate normal density estimation as an intermediate step. This is "unacceptable" at many levels.

Chapter 8

Survival Trees

8.1 Introduction

Machine Learning has its own set of terminology distinct from Statistics. In the first section, we will introduce the commonly used terms in this domain and explore them. The main topic of the chapter is by its name, survival trees. A survival tree can be obtained by multiple criteria and we will introduce them in Section 8.3 and the algorithm and construction of Survival Tree is taken up in Section 8.4. We will explore multiple options of constructing the trees and displaying them. Unlike conventional statistics, we do not have hypothesis testing metrics such as p-value, or AIC. Different criteria are required for assessing models and we look at the *variable importance* of the covariates in the construction of survival tree. Model selection in Statistics is related to the problem of *overtraining* in machine learning and we would prefer to work with a survival tree that is not overtrained. The nuances will be handled in the concluding Section 8.6.

8.2 Machine Learning Terminologies

Machine Learning (ML), Statistical Learning, Data Mining, Data Science, etc., are more or less the same even as subtle differences do set them apart. ML has seen contribution from the academic world of Statisticians and Data Analysts as well as Engineers. Thus, ML has a lot of new technical jargon about it. We will connect our earlier terminology with the ML ones.

We will begin with the conventions related to variables. The covariate vector $\mathbf{x}(t)$ is called as the *feature vector*. It is also known as the exogenous vector in Econometrics/Economics. The output variable is referred in ML as the *target variable* and endogenous variable in Economics. An observation made up of both the features and the target variable is called an *instance*. While it is common to denote the number of observations by n, the common ML notation is m, which will not be used here.

DOI: 10.1201/9781003306979-8

Most of the ML techniques fall under two classes—*supervised learning* and *unsupervised learning*. In supervised learning, we have a target variable which is modeled using the features. The purpose is to learn a function/mapping between the features and target variable such that the error is minimized and we are able to make predictions with reasonable accuracy. It has been observed that what is generally called as residual in statistics is referred to as error in the ML paradigm. We have been using thus far ϵ for error, unobservable variable. The residual is generally $r = \hat{Y} - Y$, and this is sometimes called as error in ML. Residual might be interpreted as an estimate of the error. However, the notion of error and probability is not alien in this field either, and we will continue to maintain the difference between error and residual.

In our context, the target variable is the pair of the survival time and the event indicator. A few practitioners rush to implement survival data within the ML domain by treating the event indicator as one of the feature. The practice is clearly erroneous. The event indicator is never a feature and it must be always treated as a part of the output as we have done earlier in the book. Importantly, diverse ML methods exist to handle survival data.

Supervised learning is also referred to as *learning with a teacher*. "Teacher" is used because the output variable serves as an important guide in creation of the models. Supervised learning encompasses all the scenarios where we have a target problem. Thus, classification and regression models are considered as examples of supervised learning. Time series analysis and survival analysis also fall under the umbrella of supervised learning.

Unsupervised learning deals with analysis when the goal is to understand the data as much as possible and we do not have any specific variable of interest. Multivariate analysis is an example where we often do not have a specific variable of interest. Cluster analysis is another example where explicit group labels are not put on the observations and the intent is to group together the observations which are similar with one another, leading to sub populations. Principal component analysis is a different kind of unsupervised learning. Why are the principal components an example of unsupervised learning? This is primarily because we were never looking out for a certain linear combination of the features to form the principal component.

In ML, we have a slew of methods that are neither supervised nor unsupervised, and in some sense, these are both. *Association Rule Mining* and Recommender System are two such examples. It is common to see on portals, such as Amazon, Flipkart, and Walmart, the message of 'the customers who brought this also brought ...'. Here, if the customer had purchased the books of ABG and ABGK, he might be recommended to purchase Fleming and Harrington (1991) classic. On the other hand, if the customer had purchased ABGK and Flemming and Harrington (1991), they might be recommended to purchase ABG. Thus, sometimes the unit is an input feature, and at other times, it is the output variable. Clearly, we have the case of a method being supervised as well as unsupervised.

We have one more example of an ML technique that does not fall within the contours of supervised and unsupervised methods. The method is *rein-*

forcement learning (RL). An example of RL is setting an algorithm whereby the machine will try to play the game by itself. For example, RL tries to imitate the human action of playing and failing the mobile games Temple Run and Atari. The more serious application of RL is in designing systems that learn the environment around them.

Suppose we have n observations. In a typical ML implementation, it is a practice not to use all of the n observations to learn the relationship between the features and the target variable. A few ML methods are not based on the Euclidean distance, or norm, and thereby we do not have a coefficient type of measure of the influence of the feature on the target. Thus, the question of significance or insignificance based on the fitted models does not arise and we might end up with the problems of *overfitting* or *overtraining*. However, it is equally vital that we safeguard ourselves against the overtraining problem. The reason for safeguarding is that even typical models that perform exceedingly well on the observations on which it is built, the performance might be poor on the unknown cases. In the context of linear regression model, the concept of R^2, or Adj-R^2, is versatile, while in the case of classification problem, it is common to look at measures such as accuracy, false positive, ROC, etc. These measures are distribution-free and can be computed on the fitted models using the fitted values and the actual values. This observation is put to full effectiveness in the assessment of ML models.

The ML spirit is to partition the dataset into two parts—*train dataset* and *valid dataset.* Though there are no hard and fast rules over the partition of the dataset into train and valid, a simple thumb rule is in order. If we *believe* we have a very large dataset, we can have 60% observations in the train partition and the rest in validation portion. When we have fewer observations, it is fair to have the split in 90:10 percentages. Intermediate number of observations might see a 70:30 split, or even 80:20. The validation partition is sometimes also called out as the *hold-out sample.* The validation data partition is sometimes called as *test data* even as some users make distinction between test, valid, and test partitions.

The ML models, denoted by M, are created using the training dataset and then predictions are obtained over the valid part. Since the valid part observations had no role to play in creation of the model(s) M, if the trained model has good generalizing ability, we expect it to predict the observations in the valid partition fairly well. It is here that we make a comparison between the metrics, such as R^2 or the Area Under Curve (AUC) as appropriate, of the train and valid partitions and take the decision over the usability of the model M. To clarify the idea, suppose that we get an R^2 of 85% in the training dataset. Now, if the model M has learned the patterns in the dataset properly, the R^2 in the validation partition hovers in the vicinity of 85%. The authors' observation is that a decline in performance in the validation partition prompts the users to discard the model, whereas a higher performance leads to acceptance of the model a have no problems if the accuracy is 95% in the valid part. The reader must ponder over this practice.

The test-valid partition is sometimes generalized into *k-fold cross valida-tion*. Here, the available data is partitioned into k-equal parts. Keeping the first part for validation purpose, the remaining blocks are used to create the model and the metric, say R^2, is evaluated on the first part, also referred to as the *validation block*. In the second step, the second block is reserved for the validation purpose, while the first and the remaining $k - 2$ blocks are used to create the model and the metric is obtained on the second part. Simi-larly, the process is repeated until the last block is held out as validation and the metric evaluated. The overall R^2 is averaged over the validation blocks. Multiple models are created and their average R^2 is obtained in the same manner. To keep the comparisons meaningful, the partitions are retained for all the models. We choose that model which has maximum R^2 over the hold-out partitions or the validation blocks. Thus, if we have m different models, $M_1, M_2, \ldots, M_m$, we create $m \times k$ number of models. This is similar to the jackknife and bootstrap principles.

It is important to note we do not have to carry out k-fold cross validation technique for all classes of models. Ensemble techniques such as random forest inherently adopt this approach. This will be seen in the next chapter.

Parameters. In the machine learning domain, the term parameter refers to a few elements of the algorithm which can be outrightly specified. For in-stance, the penalty factor, aka the Lagrangian multiplier, in ℓ_1 and ℓ_2 penalty, are specified over a grid. In many time series models and the linear regression model, Lagrangian multipliers are determined in a manner which maximizes the likelihood function. Another example of a parameter in the ML domain is the number of hidden layers and number of neurons in the layers. Typically, grid search or a 'fair' generalization is the way out for the choice of these numbers. Other examples of parameters are the choice of activation functions in neural networks and the type of kernels in support vector machine.

8.3 Decision Trees and Node Split Functions

Decision trees are simple in visual depiction and rich in the appeal and pretty straight-forward to understand. Parallel to Breiman's invention of the decision tree, Quinlan invented the classification tree which is often referred to as *C4.5 algorithm*. Another alternative development which gives similar result is widely known as *CHAID*, an acronym for Chi Squared Automatic Interaction Detection. The decision tree method is referred to as a *recursive partitioning* method too. A decision tree is a top-down display. The top-down display has become widespread. At the first split, sometimes called as the root node, a logical check is performed. If the criterion is met, we go to the left side of the flow, else to the right. The split partitions are also called as children nodes. At each stage, we carry out the test and move to the right or left based on the result. This process is repeated until we reach the terminal node which ends in a decision for the output variable. Depending on the problem on hand, the output would be a class labeling, a numeric prediction, or a lifetime. The class output is given by a *classification tree*, the numeric prediction by a *regression tree* while the lifetime prediction is provided by a *survival tree*. We will look at the simple example of a regression tree given in Figure 8.1.

Here, we have three input features X_1, X_2, X_3 and the output value is Y. In this scenario, the variables X_1, X_2, and Y are numeric variables while X_3 is a categorical variable. The first check performed here is whether the value of X_1 for the given observation is less than 25, $X_1 < 25$. If the criterion is not met, i.e., X_1 exceeds 25 the regression tree will return the prediction as $\hat{Y} = 40$. An additional check is required for those observations for which $X_1 < 25$. In such a case, we move to the left side and perform the logical test $X_2 > 2500$. For the scenario of $X_2 < 2500$, the prediction is 100, that is, $\hat{Y} = 100$. If we had moved to the right side, the next test is whether $X_3 =$ "Good". On meeting the criterion, the prediction is $\hat{Y} = 10$, else $\hat{Y} = 250$. The decision tree can be written in the form of an equation:

$$
\begin{aligned}
Y \;=\; & 40I\{X_1 \geq 25\} + 100I\{X_1 < 25, X_2 > 2500\} + \\
& 10I\{X_1 < 25, X_2 \leq 2500, X_3 = \text{``Good''}\} + \\
& 250I\{X_1 < 25, X_2 \leq 2500, X_3 \neq \text{``Good''}\}.
\end{aligned}
$$

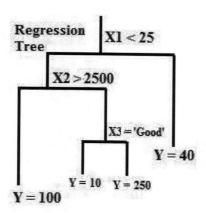

FIGURE 8.1: A Simple Regression Tree

Figure 8.1 tells us how to interpret a decision tree. We will next look at the classification problem to get a flair of the working of the decision tree. We have three blocks of display in Figure 8.2: 'A: Solvable Problem Under Partitions', 'B: A Classification Tree', and 'C: Classification Tree on the Data Display'. Here, we have two input variables X_1 and X_2. The output/target variable is a class label and the two classes are depicted on the scatterplot by a green circle and a red square. It is clear from block A in the diagram that

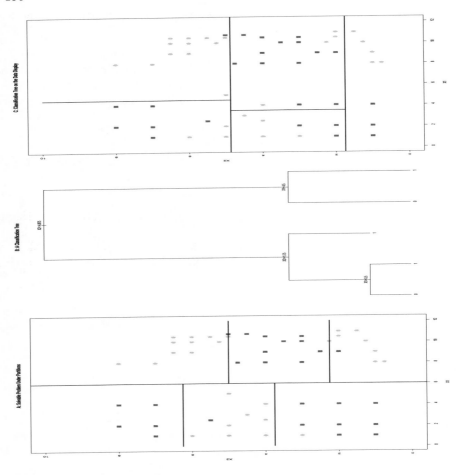

FIGURE 8.2: A Classification Tree on Display

no linear classification can achieve good accuracy. It is clear from a bird's eye view that though the labels can not be linearly separated by a single line, if we recursively partition the data, it is possible to differentiate the green circles from the red squares. For instance, a first partition can be at $X_1 = 6$. This puts the data into two blocks. Next, in the left side, two segments at $X_2 = 3.75$ and $X_2 = 6.25$ separate the green circles from the red squares. Similarly, in the right block of the data, two segments at $X_2 = 2.25$ and $X_2 = 5$ achieve the near perfect classification. We have a few misclassifications too. Indeed, it is possible to carry out further partitions and accomplish perfect accuracy. However, generalization is more important and we must not nudge the area of overtraining.

The segment drawing in part A of the figure is our idea. However, algorithms do not have such a luxury since they do not have the 'birds eye view'. Without going into the precise algorithm, the idea here is to have a peek at the bigger picture of decision trees, we will look at the classification tree obtained by the algorithm and inspect Part B of Figure 8.2. Here, the tree asks us to first check if $X_2 < 4.875$. On moving to the right side if the condition does not hold, the second check is whether $X_1 < 4.5$. If the criterion is satisfied, the observation is labeled as a red square, else as green circle. The reader can interpret the left part of the decision tree similarly. The C part of the figure are these rules implemented on the scatterplot. Of course, if we continue to partition the data, we will get higher and higher accuracy. The question is how does one carry out recursive partitioning?

The starting point of understanding decision trees is to refer to Breiman, et al. (1984)[19] and Quinlan (1993)[91]. Hastie, et al. (2009)[51] is a modern treatment while Tan, et al. (2005)[109] is an excellent treatise on various topics of data mining. We will have a look at the core algorithm of a decision tree before going into the specifics of survival tree, see Zhang and Singer (2010)[120].

The basic setup of data for construction of trees is the following. Suppose that we have n observations, and Y is the variable of interest. We note here that Y might be a numeric variable, categorical variable, or any other standard format too. The independent variables are denoted by $X_1, X_2, \ldots, X_p$. The independent variables can be numeric or categorical. How does one then split/partition the dataset? For a continuous variable, the task is a bit simpler. First, identify the unique distinct values of the numeric object. Let us say, for example, that the distinct values of a numeric object, say height in cms, are recorded as 160, 165, 170, 175, and 180. The data partitions are then obtained as follows:

- data[Height≤160,], data[Height>160,]

- data[Height≤165,], data[Height>165,]

- data[Height≤170,], data[Height>170,]

- data[Height≤175,], data[Height>175,]

The generic decision tree algorithm is given next.

ALGORITHM 8.3.1. **Generic Algorithm for Decision Tree**.

1. Identify the distinct values of the covariate X_1.

2. For each distinct value of the covariate, partition the data into the left and right parts. Find the improvisation due to the split by that value.

3. Find that covariate value which leads to maximum improvisation.

4. Repeat the previous three steps for all the variables and get that split point which gives maximum improvisation across all the variables.

5. Split the data into right and left parts according to the value and the variable identified in the previous step.

6. Repeat the previous five steps on the two parts recursively until no more splits are possible and setup the tree.

7. For a new observation, pass it down the tree until the observation reaches a terminal node and return the value of the terminal node as prediction for the observation.

For regression problems, ANOVA, or maximising the within sum of squares between partitions, is the split criterion. Let Γ denote all observations of y at a given node. Now, using a the distinct value of a covariate x_1, say x_{1d}, we partition the observed values into left part and right part. We will respectively denote the observations in the left and right part by Γ_L and Γ_R. The impurity measure at the source node is then obtained by within sum-of-squares as $i(\Gamma) = \sum_i (y_i - \bar{y})^2$. Similarly, the impurities are calculated for the left and right partitions of the data by $i(\Gamma_L)$ and $i(\Gamma_R)$. The *improvisation* is then defined by

$$\Delta(x_{1d}, \Gamma) = i(\Gamma) - i(\Gamma_L) - i(\Gamma_R). \tag{8.1}$$

The improvisation is calculated for all distinct values of the covariate x_1. We select that distinct value of x_1 for which $\Delta(x_{1d}, \Gamma)$ is the maximum. For a detailed illustration with practical data set and R code written from scratch, the reader might refer Chapter 9 of Tattar (2017)[111]. In the next section, we will discuss the development of node split functions tailored for setting up a survival tree.

8.4 Survival Tree

The choice of node split function is critical to setting up a decision tree. Algorithm 8.3.1 is general enough and depending on the split function in Step 2, the relevant class of problems is solved. Practitioners have been trying to setup survival trees right from the 1980s, for example, for censored data Ciampi, et al. (1981)[24] and Marubini, Morabito and Valsecchi (1983)[78], while for complete data, the attempt dates back to Morgan and Sonquist (1963)[86]. A comprehensive review of decision trees with survival data can be found in Bou-Hamad, et al. (2011)[14]. Chapter 9 of Zhang and Singer (2010)[120] is a brilliant beginning too. We will consider multiple choices of splits for survival data.

In the context of regression problems, the within sum of squares is maximised over the two partitions, and the values of the regressand within a node are as close as it is empirically possible. We will begin the discussion of Gordon and Olshen (1985)[48] split criterion which is explained in detail in Section 9.1 of Zhang and Singer (2010)[120].

The motivation for defining the split function in Gordon and Olshen is along the following lines. With survival data, a *node is pure* if all failure times associated with the observations in the node are equal. We then have three possible cases. In the first case, it is possible that all failures occur at the same time and the common failure time occurs after all the observations get censored. In such a case $S(t)$ will be 1 for all t. In the second case, all failure times occur at the same time, say t', and they are observed before their associated censoring times. Here $S(t)$ will be 1 up to such time t' and then becomes 0 after the failure time. In the third possible case, some failure times occur at a specific time, and rest are censored. Here, $S(t)$ will be 1 before the failure time, and it will assume a value between 0 and 1 at the failure time, and it remains at that value thereafter. Essentially, the survival curve of a pure node will have at most one drop, see Figure 9.1 of Zhang and Singer (2010)[120].

A typical Kaplan-Meier curve will have multiple jump points which can be compared with one of the three types of the pure-node curve. The non-overlapping region between these curves then serves as an assessment of the impurity of the node. Gordon and Olshen capture non-overlapping region between the curves using a generalized notion of distance function between two Kaplan-Meier survival curves defined by L^p Wasserstein metrics $d_p(., .)$. The Wasserstein distance captures the non-overlapping region between the empirical Kaplan-Meier estimator and a curve corresponding to one of the three pure node curves. However, this technique did not gain popularity in the implementation phase.

It is important to note here that Gordon and Olshen (1985)[48] clearly mention the possibility of using the likelihood ratio tests and the logrank test statistic to obtain the distance (difference/deviance) between the two children nodes, see Bou-Hamad, et al. (2011)[14], and the logrank test statistic is used in Ciampi et al. (1986)[24]. We select that covariate value by which partitioning leads to the largest value of the logrank test statistic for the samples under the two child nodes. Similarly, split functions can be setup under the assumption that the lifetimes within the child nodes follow exponential distribution, and then we apply the likelihood ratio test for comparison of equality of the exponential curves, see Davis and Anderson (1989)[35]. Therneau, Grambsch and Fleming (1990)[117] suggested use of martingale residuals from a Cox model fitted with the child nodes indicator alone, and then use the usual regression tree split function which is ANOVA.

The split criterion that will be used here is based on the full-likelihood function under the Cox regression model assuming that the hazard functions of the child nodes are proportional. This approach has been developed in LeBlanc and Crowley (1992)[70]. The improvement due to the split is captured by the *deviance* between the parent node full-likelihood function and the child node full-likelihood function. For more details, the reader can consult Subsection 9.1.3 of Zhang and Singer (2010)[120]. This method is implemented in the R package `rpart`. The `rpart` package has significant relevance.

The R package developed and maintained by Therneau has become the most popular package for creating the decision tree. Besides, Prof Therneau has significant contribution in the area of survival analysis. The `rpart` function in the package sets up classification, regression, and survival tree according as the output variable obtained is a class variable, numeric, or survival object. The user does not have to specifically mention the option of the type of tree to be created. In the next example, we will setup the survival tree and display it. To get into the specifics of the computation associated with the split function implemented in R, type and run `rpart.exp` in the console.

The algorithm for survival tree is given in the following.

ALGORITHM 8.4.1. **Generic Algorithm for Survival Tree**.

1. Identify the distinct values of covariate x_1.

2. For each distinct value of the covariate, partition the data into the left and right parts. Find the LeBlanc and Crowley deviance because of the split induced by that value.

3. Find the covariate value leading to maximum deviance.

4. Repeat the previous three steps for each of the remaining covariates. From among the p split points thus obtained, pick that split point which gives maximum deviance.

5. Split the data into right and left parts according to the value selected in the previous step.

6. Repeat the previous five steps to obtain two parts recursively until no more splits are possible to setup the survival tree.

Example 41 *A Survival Tree for the PBC Data. We continue to analyze the primary biliary cirrhosis data using the survival tree. Using the* `rpart` *object with the usual* `formula` *and* `data` *options, we setup the survival tree. We now fit the Cox proportional hazards model to make an useful observation which brings out the advantage of decision tree.*

```
> # Survival Tree
> pbc_stree <- rpart(Surv(time,status==2)~.,data=pbc[,-1])
> pbc_coxn <- coxph(Surv(time,status==2)~.,data=pbc[,-1])
> pbc_coxn$n
[1] 276
> pbc_coxn$nevent
[1] 111
> dim(pbc_coxn$y)
[1] 276   2
> dim(pbc_stree$y)
[1] 418   2
> dim(na.omit(pbc))
[1] 276  20
```

Note that we passed the same data frame through the survival tree function rpart *and to the Cox proportional hazards regression function* coxph. *The number of observations used in the Cox proportional hazards regression is* 276, *as seen by running* pbc_coxn$n, *and the number of observed events is* 111 *as given by* pbc_coxn$nevent. *The number of observations is validated by* dim(pbc_coxn$y). *But, the number of observations going into the construction of the survival tree is* 418, *see above output. Why is there this discrepancy of* 418 − 276 = 142 *observations?*

The number of observations is huge, 142, and the difference of number of observations going into the two models raises an alarm. A simple check of na.omit *shows that if we delete an observation containing a missing value for any covariate, we are left with 276 observations which are going into the Cox proportional hazards model. However, we have all observations going into the construction of the survival tree. This leads to a very important point about the way the decision trees are created. The plain square to indicate end of the example.*

Decision trees are by-and-large *non-Euclidean methods.* Here, the notion of distance is not used in terms of the vector distance. Why? This is one of the strengths of the decision trees in the way the missing data are handled by its algorithm. Recall that in maximizing the partial likelihood function, Equation 5.11, we need to calculate the term $\frac{r(\boldsymbol{\beta}, \mathbf{x}_{m_i}(T_i))}{\sum_{l \in \mathcal{R}_i} r(\boldsymbol{\beta}, \mathbf{x}_l(T_i))}$ for each observation i. Consequently, if the observation has a missing value even in one covariate, we cannot proceed further. Thus, the Cox proportional hazards regression method is compelled to shove out the observations which have missing value in even a single covariate value. Whereas, when we need to calculate the improvisation due to a split, Equation 8.1, it does not matter if the values are missing in other covariates. Thus, the decision tree makes use of all the information available in a covariate. In the example, we see that all 418 observations are used by the decision tree. This is one inherent advantage of using tree-based methods: *Decision trees are not as adversely affected by missing data as the traditional statistical methods.*

Example 42 *A Survival Tree for the PBC Data. Contd. We now look at the output produced at the console by simply running* pbc_stree *in R console. The information needs a bit of explanation. The first line is clear and it gives the number of observations that have gone into the construction of the survival tree. Next, the information of the output is arranged in terms of* node), split, n, deviance, yval, *where by* node) *gives the node number, the variable leading to the branching of the tree is given in* split, n *now becomes the number of observations available from that part of the tree downwards,* deviance *is the comparison of the full likelihood function with respect to the null model, and* yval *values need more explanation.*

Internally, the survival times are scaled exponentially and the predicted rate at the root node is 1. Thus, for the root node 1), *the* yval *value will be always 1. For the child nodes, it will be lesser than 1. Finally, the terminal nodes are indicated with the asterisk,* *, *symbol.*

```
> pbc_stree
n= 418

node), split, n, deviance, yval
      * denotes terminal node

 1) root 418 555.700 1.0000
   2) bili< 2.25 269 232.600 0.4873
     4) age< 51.24 133  76.380 0.2396
       8) alk.phos< 1776 103  30.340 0.1037 *
       9) alk.phos>=1776 30  33.090 0.6136 *
     5) age>=51.24 136 136.800 0.7744
      10) protime< 10.85 98  80.890 0.5121 *
      11) protime>=10.85 38  43.380 1.4340
        22) age< 65.38 26  24.050 0.9480
          44) bili< 0.75 8   5.189 0.3150 *
          45) bili>=0.75 18  12.550 1.3800 *
        23) age>=65.38 12   8.392 3.2680 *
   3) bili>=2.25 149 206.500 2.6970
     6) protime< 11.25 94  98.800 1.7720
      12) stage< 3.5 57  56.730 1.2620
        24) age< 43.53 25  16.660 0.6045 *
        25) age>=43.53 32  32.990 1.7990 *
      13) stage>=3.5 37  32.950 2.8310 *
     7) protime>=11.25 55  76.600 5.1840
      14) ascites< 0.5 41  52.280 4.1600
        28) age< 42.68 7   6.830 1.4340 *
        29) age>=42.68 34  37.570 5.1140 *
      15) ascites>=0.5 14  17.010 7.9060 *
```

We will next display the survival tree using the plot *function. The* plot *function displays the skeletal tree and we need to follow it up with the* text *function which will overlay with the split variables and their associated values on to the tree branches.*

```
> pdf("../Output/PBC_Stree.pdf",height=6,width=10)
> plot(pbc_stree,uniform=TRUE)
> text(pbc_stree)
> title(main="Survival Tree for PBC")
> dev.off()
```

Specifying the option of uniform=TRUE *ensures uniform spread between the nodes at the levels. When this option is set to* FALSE, *the spacings would be proportional to the error in the fit. We have 23 nodes in the output of* pbc_stree *and there are, indeed, 23 nodes seen in Figure 8.3. The output at*

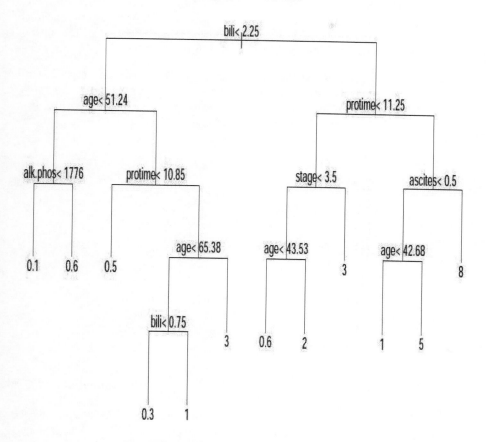

FIGURE 8.3: Survival Tree for PBC Data

the terminal node is the median survival time, which also corresponds to the constant hazard rate for the observations under that node.

 Interpretation of survival tree is straightforward and even as the reader might have already figured it out, we will consider two paths from the root node to the terminal node in Figure 8.3. If `bili<2.25`, `age>=51.24` and `protime<10.85`, then the observation will have a hazard rate of 0.5. This corresponds to node number 10) as seen in the output given two blocks above. On the other hand, if a `bili>=2.25`, `protime<11.25`, and `stage>=3.5`, the observation will have hazard rate of 3 and this corresponds to node number 13). Note that when specifying the hazard rate, we were looking at display of the survival tree given in Figure 8.3. The precise answer is reported upto four decimals accuracy in the output of `pbc_stree`. □

What is the precise improvisation in the deviance because of the given splits? Which terminal nodes do the observations belong to? How many observations are present in each of the leaf nodes? Which parameters, not in the conventional statistical sense, have gone into the construction of the survival tree? We will provide answers in continuation of the discussion of `pbc_stree`.

Example 43 ***Understanding the Survival Tree. Contd.*** *Recollect that the split covariate at root node was the bilirubin and we executed the split at the value of 2.25. What was the improvisation in the deviance because of the split? An answer to this is provided by looking out for the details by extracting* `splits` *from the fitted survival object in R.*

```
> # The Split points
> pbc_stree$splits
          count ncat  improve     index      adj
bili        418   -1 116.5655     2.250 0.00000
copper      310   -1  59.0956    85.000 0.00000
albumin     418    1  56.6642     3.315 0.00000
protime     416   -1  55.7740    11.450 0.00000

protime      40   -1   7.1586    13.050 0.00000
chol         28    1   4.6795   251.500 0.00000
copper       33   -1   3.9956    78.000 0.00000
albumin       0    1   0.8537     3.850 0.14286
> dim(pbc_stree$splits)
[1] 100    5
```

Note that we have more details here than the information provided by the 23 nodes as given earlier.

Next, we need to find how the observations end up in the leaf nodes. The 418 observations have to end up in one or other leaf node. This information is provided in the `where` *option.*

```
> # The observations in the leaf nodes
> pbc_stree$where
   1   2   3   4   5   6   7   8   9  10  11  12  13  14
  23   7  12   7  16  12   7  10  16  23  11  22   4  11
  15  16  17  18  19  20  21  22  23  24  25  26  27  28
  11   4  18  22   5  22  10  22  23   5   4  17  23  23

 393 394 395 396 397 398 399 400 401 402 403 404 405 406
   7  11   4   4   7  18  16   4  18   7  18  16   7  23
 407 408 409 410 411 412 413 414 415 416 417 418
   7   4   4   4   4   7   4  12   4   7   7   7
> # The Number of Observations at the leaf node
> table(pbc_stree$where)
```

4	5	7	10	11	12	16	17	18	21	22	23
103	30	98	8	18	12	25	32	37	7	34	14

There is a difference between leaf node and the terminal node and we leave it to the reader to figure it out.

The parameters are extracted by using $control. *To find the details regarding the control arguments, run* ?rpart.control *in R. Recollect that we did not fix specific values for the control parameters, and hence the output uses the default settings. We have criteria such as minimum split and minimum bucket. The minimum split criterion is the minimum number of observations to be available in a node before we consider it for further partitioning. The option of minimum bucket criterion corresponds to the minimum number of observations to be left in a terminal node. We have lot many options, and we will take them up for dicussion after the current example ends.*

A consequence of setting most of the parameters to zero or one, the minimalistic values, is that we end up with the problem of overfitting. In the context of a classification problem, it means that either observations of same class are present in a node, or many observations might end up having a node of their own. The problem does not simply end with overfitting because the flip side is that the tree/model loses the power of generalization. The terminal nodes must be representative of an useful pattern. For instance, if one adds patient identification number then there would be as many nodes as the number of observations which serves no purpose. It makes no sense to speak of a few observations being closer to one another because of how we serialized them.

```
> # The parameters of the Survival Tree
> pbc_stree$control
$minsplit
[1] 20
$minbucket
[1] 7
$cp
[1] 0.01
$maxcompete
[1] 4
$maxsurrogate
[1] 5
$usesurrogate
[1] 2
$surrogatestyle
[1] 0
$maxdepth
[1] 30
$xval
[1] 10
```

The   other   control   parameters   are   `cp`,   `maxcompete`, `maxsurrogate`,..., `maxdepth`. *Of course, the reader can get the details by running* `rpart.control` *in the R consol. A few of these criteria will be considered next, in Section 8.5 and the rest in Section 8.6.*   □

Issues related to prediction are also dealt with in what follows.

8.5   Prediction and Variable Importance

A survival tree can be displayed in multiple ways. Figure 8.3 begins with the root node at the top and then terminates at multiple nodes, and the constant hazard rate is reported as the value associated with the terminal node. Thus, if we run `unique(predict(pbc_stree))` in Example 42, we will have 12 unique values associated with the constant hazard rates of the terminal nodes. However, the hazard rate prediction might not be useful for a practitioner. It is a common practice, and requirement, that given the values of the covariate, the prediction return be a value on the scale of the output variable. The `rpart` created survival tree does not provide us with the prediction of the output variable. We need to use the R package `partykit`, convert the `rpart` object to a `party` object, and then use the functions and features of the `partykit` package to carry out the prediction. We will first look at the difference in the display of the survival tree as given by the `party` plots.

Example 44 *Predictions with Survival Tree.* *In the first step, we need to convert the* `rpart` *objects to* `party` *using the coercing function* `as`. *The next step is obvious, and we simply put the* `plot` *function in action. The result of the action is the plot which is reproduced here as Figure 8.4.*

```
> # Prediction and Variable Importance
> pbc_party <- as.party(pbc_stree)
> pbc_party
Model formula:
Surv(time, status == 2) ~ trt + age + sex + ascites + hepato +
    spiders + edema + bili + chol + albumin + copper + alk.phos +
    ast + trig + platelet + protime + stage

Fitted party:
[1] root
|   [2] bili < 2.25
|   |   [3] age < 51.24
|   |   |   [4] alk.phos < 1776: Inf (n = 103)
|   |   |   [5] alk.phos >= 1776: 4191 (n = 30)
```

```
|   |     [6] age >= 51.24
|   |   |     [7] protime < 10.85: Inf (n = 98)
|   |   |     [8] protime >= 10.85
|   |   |   |     [9] age < 65.38
|   |   |   |   |     [10] bili < 0.75: Inf (n = 8)
|   |   |   |   |     [11] bili >= 0.75: 3170 (n = 18)
|   |   |   |     [12] age >= 65.38: 1012 (n = 12)
|   [13] bili >= 2.25
|   |     [14] protime < 11.25
|   |   |     [15] stage < 3.5
|   |   |   |     [16] age < 43.53: 3839 (n = 25)
|   |   |   |     [17] age >= 43.53: 1746 (n = 32)
|   |   |     [18] stage >= 3.5: 1165 (n = 37)
|   |     [19] protime >= 11.25
|   |   |     [20] ascites < 0.5
|   |   |   |     [21] age < 42.68: 3428 (n = 7)
|   |   |   |     [22] age >= 42.68: 625 (n = 34)
|   |   |     [23] ascites >= 0.5: 222 (n = 14)

Number of inner nodes:    11
Number of terminal nodes: 12
> pdf("../Output/PBG_Terminal_Survival_Curves.pdf",
+     height=10,width=30)
> plot(pbc_party)
> dev.off()
RStudioGD
         1
> pbc_predict <- predict(pbc_party,type="response")
> unique(pbc_predict)
 [1]   222   Inf 1012 3839 3170   625 1165 4191 1746 3428
```

In Figure 8.3, we have the hazard rates reported in the terminal nodes. Compare the output of pbc_party with pbc_stree. The output of pbc_party as seen above is markedly different, and here it is the predicted response time, aka the lifetime, that is given for the terminal nodes. We have ten unique values at the terminal nodes, inclusive of infinity for three of them.

The output of the plot function in the above code block is given in Figure 8.4. The reader is advised to open the file PBG_Terminal_Survival _Curves.pdf available in the folder R/Chapter_09/Output. The party function uses the observations available in the terminal node and creates Kaplan-Meier curves and displays them in Figure 8.4. This gives us a fairly good view of what to expect about the lifetimes for the observations dropped at the root node and caught in one of these terminal nodes.

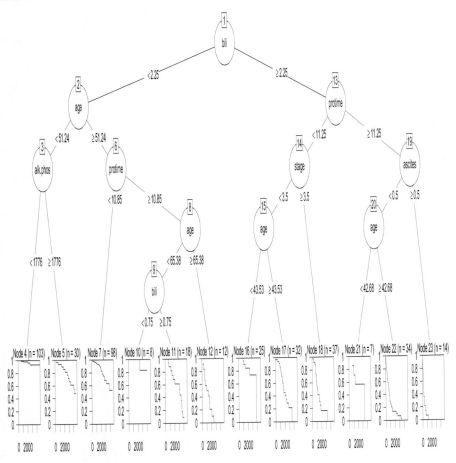

FIGURE 8.4: Survival Trees with Survival Curves in the Terminal Nodes

It can be seen from the unique values of the predictions for the survival tree that we have only nine distinct finite values for the time to death. Recall that the survival tree is essentially a non-linear regression technique. However, if the fitted values are to be closer to the actuals, the plot of the predicted vs the actuals must be linear. We first predict the time to event for the observations going into the setting up of survival tree using the `predict` *function on the* `pbc_party` *object. The* `plot` *graphical function gives us the required output and it is produced in Figure 8.5.*

```
> pbc_predict <- predict(pbc_party,type="response")
> pdf("../Output/PBC_Predictions_STree.pdf",height=10,width=10)
> plot(pbc$time,pbc_predict,xlab="Actual Time",
+      ylab="Predicted Time")
> dev.off()
```

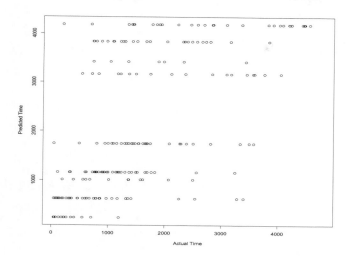

FIGURE 8.5: Predicted vs Actual Plots

Figure 8.5 does not exude confidence in the choice of the survival tree. However, it is important for us to recollect that we did not carry out this exercise earlier for the conventional statistical methods. The reader can undertake the effort on similar lines for the relative risk model and the additive regression model. A loess smoothing, a simple nonparametric smoothing technique, of the plot will provide justification for the fact that the predicted values are indeed not indicative of a poor fit.

☐

It has been mentioned in the previous section and also earlier in the chapter that the survival tree as a method is agnostic to the missing data problem. It is so on account of the nuanced details of the way a survival tree is constructed. Again, we recall, as already stated, that to select the best split by choosing the winner and then not selecting all other variables improvement is akin to throwing out the baby along with the water in the tub. In alternate words we say that it is sensible to have silver, bronze, etc. medal winners too. This is implemented in decision trees, and not just in R software or survival tree, with the use of **surrogate variables**. Before we get into that, a brief peek in the working of the survival tree is best illustrated with a small program.

Example 45 *Predictions with Survival Tree. Contd. In Figure 8.3, the first three split variables are bilirubin, age, and protime. The data object* newpbc.RData *has three observations with these three variables respectively missing in each of the observations. What happens to the prediction problem of determining the associated hazard rate or the lifetime as given by* party *function?*

```
> load("../Data/newpbc.RData")
> newpbc$bili
[1]   NA 1.1 0.5
> newpbc$age
[1] 58.77     NA 34.60
> newpbc$protime
[1] 12.2 10.6   NA
> predict(pbc_stree,newdata = newpbc)
     1      50     218
7.9063 0.5121 0.1037
```

Now, the hazard rates for the three observations have been predicted by the software as 7.9063, 0.5121, and 0.1037. How did the software handle the missing value for the three observations? Recollect from Example 43 that we had the value maxcompete *as 4,* maxsurrogate *as 5, and* usesurrogate *as 2. What is happening here is that if survival tree tries to find the value of bilirubin in the observation and it is missing, it will then search for the surrogate variable which is not depicted in the output of the survival tree. Using the surrogate variable, it proceeds further down. Similarly, for the second and third observations when age and protime values are missing, the survival tree uses the surrogate variables and proceeds to make the predictions.*

Let us turn off the option of using surrogate variables and check if the survival tree stops making the predictions. Specifying usesurrogate=0 *and slapping the* predict *function on the* newpbc *observations, we next look at the output.*

```
> pbc_tree2 <- rpart(Surv(time,status==2)~.,data=pbc[,-1],
+     usesurrogate=0)
> predict(pbc_tree2,newdata = newpbc)
     1      50     218
1.0000 0.4873 0.1053
```

The survival tree steadfastly gives the outputs not withstanding the missing values. However, the predicted output values are different from the earlier values of 7.9063, 0.5121, and 0.1037. How does the survival tree work here? The hint probably lies in the fact that when the bilirubin value is missing, the predicted hazard rate is the unit value, which is that of the root node. Thus, when the observations can not be passed down the tree, the survival tree returns the hazard rate of the current node as the prediction. This holds true whether we have a classification tree or a regression tree.

□

Note that the prediction computation does is not violate the basic principles as one might be tempted to conclude here. It is not the case that partial evaluation is happening in that if the first value x_1 is missing, we simply calculate, say, $\beta_2 x_2 + \ldots + \beta_p x_p$. The prediction is based on all the Y values available in the node at that position. This explanation is provided here to clarify predictions are obtained with survival tree.

While it is true that we do not have p-value type of metrics in aiding our call to declare a variable as significant or otherwise, the improvisations effected by the variables at the split nodes are an indication of their importance. After all, if a variable is useful in the construction of the survival tree, the variables must turn up as the winners at the tree splits. The more improvisations get affected by the variable, the better it is. This aggregation is accumulated in the metric that is popularly known in the decision tree literature as *variable importance*. For a clear and computational explanation of the concept, refer pages 100–104 of Tattar (2018)[112]. For any `rpart` tree, the variable importance is calculated. It is easily obtained by the extractor `$variable.importance`.

Example 46 *Understanding the Variable Importance in a Survival Tree. We obtain the variable importance for the* `pbc_stree` *in the next block. The bar plot gives the relative importance of the variables.*

```
> as.data.frame(pbc_stree$variable.importance)
          pbc_stree$variable.importance
bili                            138.008
protime                          70.867
age                              54.548
albumin                          32.240
edema                            25.576
stage                            15.231
ascites                          14.094
alk.phos                         13.441
platelet                         10.018
sex                               2.453
copper                            2.115
ast                               1.692
> pdf("../Output/Variable_Importance_PBC.pdf",height=20,width=20)
> barplot(sort(pbc_stree$variable.importance,decreasing = F),
+     horiz=TRUE)
> dev.off()
```

It is clear from the output that the best five variables in terms of importance are bilirubin, protime, age, albumin, and edema. Besides the "Top X" number of features, how can we choose the most number of useful features? The barplot can be interpreted in similar way as the scree plot of the principal component analysis where standing on top of the highest eigen value bar, we look down and try to find the debris after which we can ignore further variables. Of course, this introduces subjectiveness of the analyst. In Figure 8.6, a few might treat the debris begins from sex covariate downwards, while others might vouch for stage covariate downwards.

□

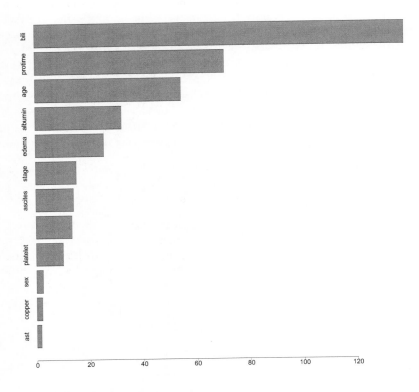

FIGURE 8.6: Variable Importance of PBC Survival Trees

In the concluding section, we will look at a few more nuanced aspects of survival tree.

8.6   Understanding Different Facets of Survival Tree

In the creation of the survival tree `pbc_stree`, we did not specify any specific parameter and went ahead with the default settings. We had earlier remarked that overfitting is often a concern in machine learning models. We now change the values of minimum split `minsplit` and minimum bucket `minbucket` values to 40 and 12 from the earlier values of 20 and 7 respectively. We need to have at least 10% of original number of observations before we effect a split. Further, we do not want to end with any node which has less than 3% observations. This will ensure that we are not overtraining the survival tree.

The parameter cp stands for complexity parameter. In the regression problem, we have within sum of squares, or ANOVA, as the split criterion, and impurity in the context of the classification problem. Here, we have deviance for the construction of survival trees. Presume that for a given data, the deviance is 1000, and we specify the complexity parameter to be 0.2. This translates into the algorithm that in the next split, the deviance should be at most 800. Otherwise, such splits are not carried out. We use the specification of cp = 0.03 to ensure that we do not have too very small splits.

Example 47 *Pruning Survival Trees with Minimum Splits, Observations, and Complexity Parameter. Using the* minsplit *and* minbucket *options, we first setup a survival tree in* pbc_stree_MO_MN, *and then specifying* cp=0.03, *we create* pbc_stree_CP. *We plot the two survival trees side-by-side to facilitate comparison.*

```
> # Pruning the Tree
> # Minimum observations at time of split be 10%
> # and minimum number of observations in terminal node be 3%
> pbc_stree_MO_MN <- rpart(Surv(time,status==2)~.,data=pbc[,-1],
+                           minsplit=40,minbucket=12)
> pdf("../Output/Pruned_Trees.pdf",height=10,width=20)
> par(mfrow=c(1,2))
> plot(pbc_stree_MO_MN,uniform=TRUE)
> text(pbc_stree_MO_MN)
> # Pruning with Complexity Parameter
> pbc_stree$cptable
         CP nsplit rel error xerror    xstd
1  0.20971      0    1.0000 1.0044 0.04760
2  0.05601      1    0.7903 0.8110 0.04475
3  0.03500      2    0.7343 0.8242 0.04651
4  0.02329      3    0.6993 0.8351 0.05059
5  0.02255      4    0.6760 0.8412 0.05249
6  0.01969      5    0.6534 0.8570 0.05562
7  0.01641      6    0.6337 0.8514 0.05637
8  0.01367      7    0.6173 0.8369 0.05506
9  0.01276      9    0.5900 0.8141 0.05453
10 0.01135     10    0.5772 0.8165 0.05491
11 0.01000     11    0.5659 0.8118 0.05495
> pbc_stree_CP <- rpart(Surv(time,status==2)~.,data=pbc[,-1],
+                         cp=0.03)
> plot(pbc_stree_CP,uniform=TRUE)
> text(pbc_stree_CP)
> pbc_stree_CP$cptable
       CP nsplit rel error xerror    xstd
1 0.20971      0    1.0000 1.0032 0.04759
2 0.05601      1    0.7903 0.8334 0.04558
```

```
3 0.03500        2     0.7343 0.8160 0.04523
4 0.03000        3     0.6993 0.8071 0.04689
> dev.off()
```

Figure 8.7 gives the two pruned survival trees. It is important for the practitioners to do all sorts of diagnostics and fine-tuning before arriving at a finished and useful survival tree.

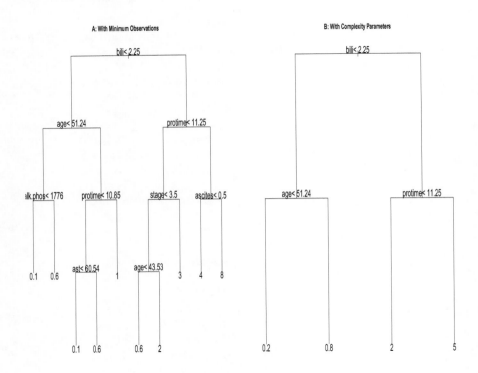

FIGURE 8.7: Pruned Survival Trees

Survival trees, as with any other statistical technique, in default settings will seldom meet the objective of the problems laid out. Modeling is a continuous iterative process and it is not different for ML technique. The topics deliberated here provide us lot of options which help in improving the analysis.

8.7 Exercises

Exercise 8.1 *Create a survival tree for the veteran's lung cancer study, see Exercise 4.3, with default settings. Obtain the variable importance of the covariates and compare it with the significant variables identified by the fitted Cox proportional hazards model.* □

Exercise 8.2 *Express yourself how you would go about finding the parameters to be pruned for the survival tree fitted in the previous exercise for the veteran's lung cancer study.* □

Exercise 8.3 *In Example 35, we saw that the* `coxph` *method issued a warning* `Ran out of iterations and did not converge`. *Create a survival tree for* `Surv(time,event)` *with the same set of covariates and record your observation whether you get any sort of warning message. If there is no such warning as in the Cox PH model case, argue out why that is so?* □

Exercise 8.4 *In the* `rpart` *implementation, the method for construction of survival tree, find more details by running* `?rpart` *in the R console, the split criterion is mentioned as* `If 'y' is a survival object, then 'method = "exp"'` *is assumed. Figure out the the split criterion in the* `party` *package.* □

Exercise 8.5 *Increase the complexity parameter* `cp` *from the default value of 0.01 to 0.05, 0.1, and 0.2 and create survival trees for any dataset of your choice. Do you expect the tree size to increase or decrease? Is your intuition matched by the plots of the survival trees for the different complexity parameter values.* □

8.8 Más Lejos Temas

Decision trees have been developed by machine learning community independently of statisticians, see Quinlan (1983)[91]. It is thus not surprising to see variety of options in splitting the nodes for survival data. The terminal nodes of the survival trees consist of the Kaplan-Meier curves. Practitioners who are well versed with the theory underlying Kaplan-Meier curves and their interpretation will find survival trees quite a useful technique while someone taking it as yet another technique may find the interpretation a daunting task. A simple workaround is to generate the pseudo-observations and then fit widely popular regression tree. This should be strictly be used as a last resort.

Survival Analysis

Trees for multi-state models have not yet been created. However, Tattar and Vaman (2014)[115] have provided tests for testing statistical hypothesis that the transition probability matrices of two populations are equal. Using this test, we can then split the multi-state model into two partitions and also setup the survival tree for the clinical studies with multi-state data.

Chapter 9

Ensemble Survival Analysis

9.1 Introduction

A fully grown survival tree is one with minimum parametric specification. The fully grown tree would have high variance because of its pursuit to eliminate bias. It is well known that in general the decision trees are prone to overfitting, and they are in general useful when we have too many categorical variables as the explanatory variables. Recall that the decision tree in general, and survival tree in particular, is a fitting techniques which is non-Euclidean in nature. The pruning aspect notwithstanding, we will continue to have the high variance problem for decision trees in our applications. Thus, we need techniques that will help us overcome the problem of high variance.

In the context of the general decision tree, Breiman (1996)[16] introduced the concept of *bagging*. Bagging is the blending of the two words—bootstrap and aggregation. As with many popular acronyms, like LASSO, many would not know that bagging indeed means bootstrap aggregation. The way bagging works is that a tree is setup on the bootstrap sample of the observations, and then we obtain a large number of trees, say B. Prediction is then carried out for those observations which are left out of the bootstrap sample. Remember that in a bootstrap sample, we have about 63% distinct observations getting selected because of sampling with replacement. The prediction is then carried out for the remaining 37% observations, and the aggregation is done over the available number of predictions. Here, we have an ensemble of B number of survival trees. What is an ensemble? This is the topic of the next section.

Even as a bootstrap sample is obtained for setting up a survival tree in the bagging algorithm, it turns out that in practical terms, the same set of features/covariates turn out to be important from the notion of "variable importance". An improvisation is therefore required in order to reduce the variance and also to give other features a chance to be more significant in terms of predicting the output variable. Breiman (2001)[17] invented an important technique called as *random forests*. Here, at each time we attempt a split, a certain number of covariates are selected to be potential candidates for partitioning the dataset. Because we have a random selection of the covariates, each tree is likely to have a different set of variables being selected. Thus the

DOI: 10.1201/9781003306979-9

word "random" is an integral part of the technique's name. Random forest is an extension of the bagging method.

A third pathbreaking technique in Machine Learning is known as *boosting*. The required number of trees in bagging and random forests technique can be parallelly obtained and their aggregation executed suitably. Boosting is a sequential method which works in a specific order and it converts the poor learners into stronger ones. In the original work, the technique was invented by Freund and Schapire (1996)[45] for the classification problem. Later this technique has been extended to a host of other problems. We will apply this technique to the Cox proportional hazards model beginning with a peek into the general ensemble learning.

9.2 Principle of Ensemble Learning

Ensemble learning takes into account all the models which are effective by a minimum level of accuracy criterion and combines them together to produce an overall 'most effective' model, say ensembled model. Two imminent questions arise naturally. Is it indeed possible to ensure that the ensemble model is better than each of the candidate models? The second question relates to whether the ensemble model is robust? The answer to both the questions is positive. Before we get into further details, a brief detour is in order.

In the popular television show 'Who Wants to Be a Millionaire?', multiple lifelines are provided. A host of similar television shows has wide viewership. Among the lifelines, we are interested in two of them— public opinion and expert support. The experts are generally authoritative persons in general affairs and it might be reasonably said that their answers are accurate on 90% or more occasions. In case of public opinion, it is an ensemble of people who buy tickets to enjoy the recording of the show and they vote one of the four objectives to the best of their abilities. In many cases, the public are divided significantly between two and three options because they are themselves not sure of the correct answer. More often than not, the contestant on the hot seat of the show goes with the majority vote in spite of being aware that quite a few occasions exist when the majority vote was a wrong answer too. In a casual survey, the authors have found that many people expect the expert to be hands-down more accurate than the public opinion, and it also reflects in the choice of the candidates who turn to the experts before seeking out the public opinion. Defying the odds, almost in literal sense, it turns out that the public opinion, the majority vote, is empirically more accurate than the expert. How did the tables turn? Probabilistically speaking, this is not surprising. If each of the voter of the four alternatives to the question is 50% or more accurate in the current affairs, which almost everybody is, with an ensemble of 20 or more voters, the majority vote will be accurate on more than 90% of the occassions.

Dataset/ Model	Hypothyroid	Waveform	German	Iris	Pima Indian Diabetes
Neural Network	98.27%	88.40%	72.52%	100.00%	67.32%
Logistic Regression	97.33%	88.73%	75.72%	100.00%	75.10%
Naïve Bayes	97.33%	86.01%	80.83%	100.00%	78.21%
Decision Tree	98.74%	84.35%	70.61%	100.00%	75.88%
SVM	98.43%	91.71%	75.40%	100.00%	76.65%

FIGURE 9.1: Accuracy of ML Models

It is also noted here that generally people who are less confident of the answer indeed refrain from voting. In fact, this aspect of the public opinion is central to the success. The majority voting takes a beating if less accurately informed voters participate in the opinion.

The Sections titled 'The right model dilemma!' and 'An ensemble purview' in Chapter 1 of Tattar (2018)[112] brings out the issues of sticking with a single winner model. Figure 9.1, which is Table 1 of Chapter 1 of Tattar (2018)[112], gives the accuracy of five classification models over five different datasets. We have five datasets here: hypothyroid, waveform, German credit, iris, and pima Indian diabetes. The five classification models used are neural networks, logistic regression, naive Bayes, decision tree, and support vector machines. It can be seen, unsurprisingly, that we do not have a single class of models that is better than the other models for all datasets. In the 'An ensemble purview' section, Tattar (2018)[112] illustrates for the German credit dataset that though the naive Bayes method is most accurate for the classification problem, it is the ensemble prediction that gives the maximum accuracy by the Area Under Curve (AUC) criterion.

We have two types of ensemble—homogeneous ensemble and heterogeneous ensemble. In the homogeneous ensemble technique, all models are members of the same machine learning family. For instance, all homogeneous ensemble models will be variants of neural networks with the models differing in the number of neurons, hidden layers, and activation functions. In the heterogeneous ensemble, we mix all types of models so long as they satisfy the inclusion criteria, such as minimum 50% accuracy. Here, the models might encompass neural networks, decision trees, support vector machines, rule-based system, etc. Sometimes, ensemble of ensembles is also considered in the sense that an ensemble method like bagging or boosting might go as a single model in a larger ensemble.

In this chapter, we will be developing homogeneous ensemble only. Bagging, random forests, and boosting are all examples of homogeneous ensemble. While the first two are based on the survival trees as the core learners, boosting will be applied to the Cox proportional hazards model. What matters is the choice of models that we are dealing for this kind of ensemble.

Heterogeneous ensemble is indeed a boon in the Machine Learning domain. The primary reason for this statement is that so long as the base model is decently accurate, we will not have to really bother about selecting one over another. This is especially true when we are dealing with the prediction problem. We have learned a host of techniques in Part I Classical Survival Analysis. In the context of regression problems, we have seen Miller's method, Buckley-James estimator, Koul-Susarla-van Ryzin, Cox proportional hazards, and Aalen's additive hazards model. Whenever the models meet the minimum accuracy criterion, we do not have to worry about selecting them and instead put all of them in action for prediction in future cases. The prediction would be a much robust solution. The heterogeneous ensemble has a slight advantage over the homogeneous ensemble.

We do not know of any analytical method that is free of assumptions. Assumptions are basic and a fundamental assumption in ensemble is that all the models are independent of each other. In the case of homogeneous ensemble, the bootstrap technique ensures that the models are based on different samples of the original dataset. It does not, however, imply that the different models are independent of one another. Model independence is more difficult to establish. In the context of heterogeneous ensemble, it is conceivable that the models would not be identical. This is one advantage of the heterogeneous ensemble. We note that there still is a long list of open problems in ensemble learning, and more so for ensemble survival analysis.

9.3 Bagging Survival Trees

As mentioned in the introductory section, bagging stands for bootstrap aggregation as a technique to reduce the variability of the decision trees which are notorious for having large variance. Hothorn, et al. (2004)[55] developed the bagging algorithm for survival trees.

We next use the survival tree algorithm to carry out the bagging algorithm.

ALGORITHM 9.3.1. **Generic Algorithm for Bagging Survival Tree**.

1. Draw a random sample of size n with replacement from the data consisting of n observations. The selected random sample is called a *bootstrap sample*.

2. Construct a survival tree for the bootstrap sample using Algorithm 1.

3. Do not prune the tree.

4. Carry out the desired prediction for the observation. Choose an out-of-bag prediction if prediction is of primary concern.

5. Repeat steps 1–4 a large number of times, say B.

6. For each observation aggregate the predictions according to the requirement-average for regression and majority vote for classification problem.

The R package `ipred` can carry out bagging for survival trees. Here, we will use a different package for this purpose by tweaking the number of covariates for selection. We will continue by specifying the number of available covariates in the option `mtry` of the `rfsrc` function available in `randomForestSRC` package. With the change in specification, we are able to implement Algorithm 1. Details of the `randomForestSRC` package will follow in the next section.

Example 48 *Bagging Survival Trees for the PBC Data. We will now try to model the survival times of the PBC dataset using all the covariates and specifying* `mtry = 17` *in the options. Here, we setup the bagging algorithm for 500 trees with the* `ntree` *specification. When we turn on the option of tree error rate in* `tree.err = TRUE`*, we are asking the function to save the error rates of each tree. This is importantly required in other computations.*

```
> pbc_bagging <- rfsrc(Surv(time,status==2)~trt + age + sex +
+                      ascites + hepato + spiders + edema + bili +
+                      chol + albumin + copper + alk.phos + ast +
+                      trig + platelet + protime +
+                      stage, ntree=500, tree.err = TRUE, mtry = 17,
+                      pbc)
> pbc_bagging$splitrule
[1] "logrankCR"
> pbc_bagging$nodesize
[1] 15
> pbc_bagging$ensemble
[1] "all"
> vimp(pbc_bagging)$importance
            event.1     event.2
trt       7.365e-05  -0.0001031
age       5.380e-02   0.0157558
sex       1.602e-03  -0.0001552

platelet -6.034e-04  -0.0001569
protime   7.882e-03   0.0024419
stage     3.876e-03   0.0013788
```

Specification of formula and other details for bagging have been already explained. We have used here `logrankCR` *for the splitting purpose, and the node size is restricted at 15. All observations are used for assessing the performance of the method. To avoid the problem of overfitting, we can use the option of out-of-bag fitting. The variable importance in each of the 500 trees is aggregated as shown in the R output above. Now, it is critical to compare the variable importance obtained here with that for the single survival tree as seen in Example 46. However, we have variable importance for two events here while we had single column in the previous chapter. The* `rfsrc` *obtains the values for the failure and censored times. This also holds true for prediction purpose where we will get two types of lifetimes: one for the failure times and another for censored one.*

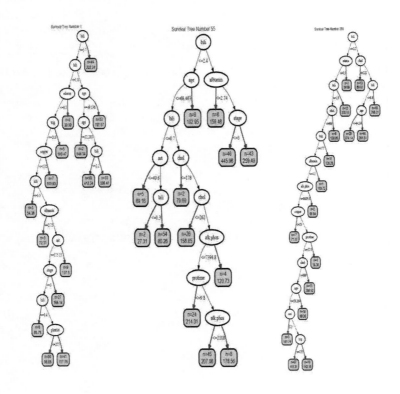

FIGURE 9.2: Bagging Survival Trees

```
> head(pbc_bagging$err.rate,5)
     event.1 event.2
[1,]   0.3304  0.2309
[2,]   0.2456  0.2448
```

```
[3,]   0.2421   0.2304
[4,]   0.3179   0.2267
[5,]   0.3231   0.2106
> tail(pbc_bagging$err.rate,10)
        event.1 event.2
[491,]   0.2229   0.1724
[492,]   0.2225   0.1724
[493,]   0.2192   0.1718
[494,]   0.2192   0.1719
[495,]   0.2192   0.1721
[496,]   0.2197   0.1721
[497,]   0.2197   0.1723
[498,]   0.2192   0.1721
[499,]   0.2197   0.1721
[500,]   0.2192   0.1719
> pdf("../Output/PBC_Bagging_Error_Convergence.pdf",
+      height=10,width=15)
> plot(pbc_bagging,plots.one.page = TRUE)
> dev.off()
```

In setting up pbc_bagging, *we specified the number of trees in the bagging algorithm at five hundred with* ntree = 500. *What was the basis for selecting the number? Is that a number closer to the number of observations or a multiple of the number of covariates going into the model? The minimum sample number is not known in most statistical problems either, especially in the context of regression problems. We are typically off in a worse way in terms of the mathematical clarity with the machine learning methods. Nevertheless, we compute the error rate with respect to the two events of observing death and censoring. Thus, as the number of trees increases, we require the error-rates to stabilize, and not necessarily the ideal zero error-rate, so that we can conclude having put enough number of trees in the bagging method. The error rates plot is easily obtained, and we use it on the bagging object to obtain Figure 9.3 which shows that bagging method has saturated after around 300 number of trees. It is important to note that the number of trees is not related to the number of observations at all.*

The tree display is one of the simplest and easily interpretable technique and hence the popularity of the tree-based methods. We have seen the survival trees on multiple occasions and have said repeatedly here that bagging is an ensemble of survival trees. Can we then see a few of the five hundred trees that go into the making of the pbc_bagging? *The* get.tree *function can be applied on the fitted bagging object to obtain the plot of a specific survival tree.*

```
> plot(get.tree(pbc_bagging,1),
+    title=paste("Survival Tree Number",1))
> plot(get.tree(pbc_bagging,55),
+    title=paste("Survival Tree Number",55))
```

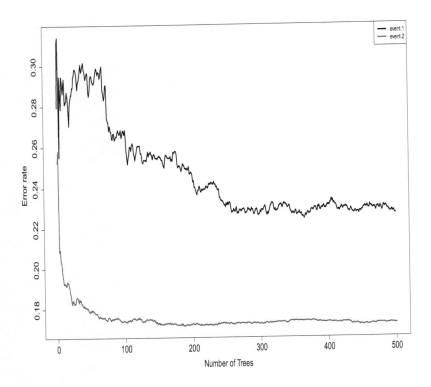

FIGURE 9.3: Bagging Convergence Rate

```
> plot(get.tree(pbc_bagging,288),
+    title=paste("Survival Tree Number",288))
> class(get.tree(pbc_bagging,288))
[1] "Node" "R6"
```

The result of the above code is produced in Figure 9.2. A remark is in order. The program is run in an R Studio in Ubuntu OS. Each of the plot above with the specified tree number is displayed in the 'Viewer' panel. If the codes are run in base R version, the gnome terminal in Ubuntu and basic R in Windows, the output is produced on the default internet browser. The authors' have copied the output of the viewer file, and edited it using a suitable graphical device. The reader should not expect a single output as shown in Figure 9.2. Why is it complex? The output of the plot is not put in the regular R graphical devices, and this can be evaluated as shown by the class *function on the tree extracted from the* pbc_bagging *object. The class of* Node *and* R6 *cannot be ported to a graphical device. Thus, it is not possible to extract and save the output to pdf or jpeg files as we have been doing earlier. This is to the best of the authors' knowledge and there might be better ways of producing the output of the tree displayed in cleaner and efficient ways.*

□

While it is not possible to look at statistical significance through p-values, AIC, etc. with trees, we can still attempt to make an effort towards finding the significant variables. With ensemble of survival trees, there are three methods of finding the significant variables—(i) minimal depth, (ii) variable hunting, and (iii) variable hunting with variable importance. The minimal depth method depends on the concept of the maximal subtree. What is a maximal subtree though?

It is our intuition that the closer a variable is to the root node, significant is the variable since it leads to maximum improvisation. The maximal subtree for a variable X is the largest subtree whose root node splits at a value of x. Note that the number of maximal subtrees can be one or more for a variable. For the `pbc_bagging`, we get the top variables by the maximal subtree as follows:

```
> max.subtree(pbc_bagging)$topvars
[1] "age"      "edema"    "bili"     "chol"      "albumin" "copper"
```

We will carry out the variable selection by each of these methods in the next example.

Example 49 *Variable Selection in Bagging. We will begin the variable selection using the minimal depth method first, and then check out the best variable using the **variable hunting** and **variable hunting with variable importance** methods. The first method is the quickest for finding the best variables for the PBC dataset. The other two methods based on variable hunting take much longer time to execute as they are based on the variable importance technique.*

```
> var.select(pbc_bagging, method = "md")
minimal depth variable selection ...
----------------------------------------------------------
family            : surv-CR
var. selection    : Minimal Depth
conservativeness  : medium
x-weighting used? : TRUE
dimension         : 17
sample size       : 276
ntree             : 500
nsplit            : 10
mtry              : 17
nodesize          : 15
refitted forest   : FALSE
```

```
model size        : 6
depth threshold   : 5.4
PE (true OOB)     : 21.92 17.19

Top variables:
        depth vimp
bili    0.464  NA
age     3.612  NA
edema   4.018  NA
albumin 4.782  NA
chol    5.008  NA
copper  5.030  NA
------------------------------------------------------------
```

The minimal depth method does not take much time to compute. On the other hand, the variable hunting and variable hunting with variable importance methods require simulation and hence take a lot of time to execute. On laptops with moderate speed, the two methods require around ten to thirty minutes each to execute. Nevertheless, if we have to select the best variables using the minimal depth method, the variables would be bilirubin, age, edema, albumin, cholesterol, and copper. The reader is encouraged to run the programs with these six variables alone and decide if the performance deteriorates. We next run the variable selection using the other two methods.

```
> var.select(pbc_bagging, method = "vh")
-------------------- Iteration: 1  --------------------
 selecting variables using Variable Hunting ...
 PE: 76.52      dim: 1
-------------------- Iteration: 2  --------------------
 selecting variables using Variable Hunting ...
 PE: 37.1       dim: 3
-------------------- Iteration: 3  --------------------
 selecting variables using Variable Hunting ...
 PE: 8.421      dim: 3

-------------------- Iteration: 48  --------------------
 selecting variables using Variable Hunting ...
 PE: 25.84      dim: 4
-------------------- Iteration: 49  --------------------
 selecting variables using Variable Hunting ...
 PE: 26.19      dim: 1
-------------------- Iteration: 50  --------------------
 selecting variables using Variable Hunting ...
 PE: 22.22      dim: 4
fitting forests to final selected variables ...
```

```
------------------------------------------------------------
family              : surv-CR
var. selection      : Variable Hunting
conservativeness    : medium
dimension           : 17
sample size         : 276
K-fold              : 5
no. reps            : 50
nstep               : 1
ntree               : 500
nsplit              : 10
mvars               : 4
nodesize            : 2
refitted forest     : TRUE
depth ratio         : 0.0996
model size          : 2.94 +/- 1.284
PE (K-fold)         : 38.17 +/- 18.74

Top variables:
         depth rel.freq
trt       1.279     22
hepato    1.237     22
alk.phos  1.304     22
------------------------------------------------------------
> var.select(pbc_bagging, method = "vh.vimp", nrep = 50)
-------------------- Iteration: 1  --------------------
 selecting variables using Variable Hunting (VIMP) ...
 PE: 52.31     dim: 1
-------------------- Iteration: 2  --------------------
 selecting variables using Variable Hunting (VIMP) ...
 PE: 44.59     dim: 3
-------------------- Iteration: 3  --------------------
 selecting variables using Variable Hunting (VIMP) ...
 PE: 44.9      dim: 2

-------------------- Iteration: 49  --------------------
 selecting variables using Variable Hunting (VIMP) ...
 PE: 38.46     dim: 3
-------------------- Iteration: 50  --------------------
 selecting variables using Variable Hunting (VIMP) ...
 PE: 18.42     dim: 2
fitting forests to final selected variables ...
------------------------------------------------------------
family              : surv-CR
var. selection      : Variable Hunting (VIMP)
```

```
conservativeness     : medium
dimension            : 17
sample size          : 276
K-fold               : 5
no. reps             : 50
nstep                : 1
ntree                : 500
nsplit               : 10
mvars                : 4
nodesize             : 2
refitted forest      : TRUE
model size           : 2.12 +/- 0.7183
PE (K-fold)          : 43.1 +/- 21.89

Top variables:
        rel.freq
protime      28
hepato       22
stage        20
----------------------------------------------------------
```

The simple variable hunting selection method returns treatment, presence of hepatomegaly, and histologic stage of disease as three important variables. The variable hunting with variable importance method gives out the protime, hepatomegaly, and the histoloic stages are important variables.

We will close the discussion of bagging application to the PBC dataset with the prediction problem. As with the error rate for the two types of event, we obtain the predictions for both the events.

```
> # Prediction with Bagging
> pbc_pred <- predict(pbc_bagging)
> head(cbind(pbc_pred$predicted,pbc$time,pbc$status))
      event.1 event.2
[1,]    10.22  3555.1  400 2
[2,]   114.21   476.3 4500 0
[3,]    76.36  1790.0 1012 2
[4,]   111.93  1487.6 1925 2
[5,]   825.58  1235.5 1504 1
[6,]   106.06   534.2 2503 2
Warning message:
In cbind(pbc_pred$predicted, pbc$time, pbc$status) :
  number of rows of result is not a multiple of
  vector length (arg 2)
> load("../Data/newpbc.RData")
> new_pred <- predict(pbc_bagging,newdata = newpbc,
+    na.action ="na.impute")
```

```
> new_pred$predicted
      event.1 event.2
[1,]    21.05  3346.0
[2,]   199.74   553.8
[3,]   106.36   235.3
```

Unlike in the regression problem, we can not obtain an evaluation metric like mean absolute percentage error. Evaluation is possible using Brier metric and we leave it to the reader to figure out the details.

□

We next move to the important ensemble method of random forests.

9.4 Random Survival Forests

The bagging ensemble reduces the variance of a survival tree. The algorithm discussed in Section 9.3 has been specifically for survival data. Figure 9.2 shows three out of the five hundred survival trees. We have intentionally used the rfsrc function from the randomForestSRC package. There is another excellent R package ipred which can perform bagging of survival data and the convenience offered is that we can save each tree of the bagging to a pdf file. Using the R function bagging from the package, we will save each tree to the file PBC_Bagging2_Trees.pdf. It would be available in the R code bundle's folder Chapter_10/Output. The following R code accomplishes the required objective:

```
> library(ipred)
> pbc_bag2 <- bagging(Surv(time,status==2)~trt + age + sex +
+    ascites + hepato + spiders + edema + bili + chol + albumin+
+    copper + alk.phos + ast +trig + platelet + protime+
+    stage, nbagg=500,keepX=TRUE,coob=FALSE,pbc[,-1])
> pdf("../Output/PBC_Bagging2_Trees.pdf")
> for(i in 1:500){
+    tt <- pbc_bag2$mtrees[[i]]
+    plot(tt$btree)
+    text(tt$btree,use.n=TRUE)
+    }
> dev.off()
null device
          1
```

The reader should open the file PBC_Bagging2_Trees.pdf and give a cursory glance over at least a hundred trees of the bagging solution. The first impression of the authors, it would be subjective as we understand, is that bilirubin and age prominently dominate the first three splits of the trees. Moreover, since there is enough overall variation across the trees, an impression might be formed that a few patterns of the tree shape sweep across hundreds of trees. This gives the overall impression that bagging might throw up the same set of variables as significant, and that once a few patterns are zeroed on, there might be a repetition of them. Thus, while the variance is reduced, the bias of selecting the same set of features remains prominent.

Breiman (2001)[17] came up with an innovative modification of the bagging algorithm which overcomes the problem of the same set of features dominating a tree early in the construction. The brilliant turn over occurs at the time of splitting the data into two partitions. Breiman suggested that instead of selecting the split among all the variables, a random number of covariates/features will be considered as potential candidates for splitting. The general recommendation for considering the number of features, assuming p of them, at a split point is $[p/3]$ for classification problem, and $[\sqrt{p}]$ for other classes, where $[.]$ denotes the integer part of the argument. By restricting the number of features at every split point, we expect to overcome the problem of overtraining. Since we consider random features to split, the trees under this algorithm, generally, have different behaviour and shape and hence the ensemble of such decision trees is referred to as *random forests*. Next, we give the algorithm of the random forest for the generic case.

ALGORITHM 9.4.1. **Generic Algorithm for Random Survival Forest**.

1. Draw a random sample of size n with replacement from the data consisting of n observations. The selected random sample is called a *bootstrap sample*.

2. Construct a survival tree for the bootstrap sample using Algorithm 1 with the following caveat:

    (a) At each split of the tree, consider only a random number of features.

    (b) Do not prune the tree.

3. Carry out the desired prediction for the observation—choose an out-of-bag prediction option if prediction is of primary concern.

4. Repeat steps 1–3 a large number of times, for example, B.

5. Aggregate each observation according to the requirement.

We note here that a few practitioners, as well as software experts skip the condition of pruning and proceed with it. It is not a stumbling block though.

Ishwaran, et al. (2008)[57] develop the random survival forest. The logrank test is the criterion for splitting nodes in the setting up of the tree. Construction of random forests based on the bootstrap sample poses several challenges. For instance, we need to ensure that at each split, the number of patients for whom the event is observed meets a minimal threshold. Ishwaran, et al.'s paper consists of host of a other crucial topics using the forests—ensemble cumulative hazard, ensemble mortality, etc.

The R function `rfsrc` basically stands for random forests for survival, regression, and classification data. We intentionaly used this function for bagging with the option `mtry = 17` which forced the implementation to consider all covariates at every split point and thereby implementing the bagging algorithm. Assuming familiarity of coding for error rate convergence, extracting trees of the ensemble/forest, variable importance extraction and the three methods of variable selection, we directly jump to random forest analysis. Since we do not specify `rfsrc` with the `mtry` option, implementation will choose $\sqrt{17} \approx 4$ randomly selected features each time it takes up the splitting problem.

Example 50 *Random Survival Forests for the PBC Data. Using the* `rfsrc` *function in the usual way without specifying* `mtry`*, we repeat the appropriate analysis in the next code block.*

```
> pbc_RSF <- rfsrc(Surv(time,status==2)~trt + age + sex +
+     ascites + hepato + spiders + edema + bili + chol + albumin+
+     copper + alk.phos + ast +trig + platelet + protime+
+     stage, ntree=500, tree.err = TRUE,
+     block.size = 1, pbc[,-1])
> pbc_RSF$splitrule
[1] "logrankCR"
> pbc_RSF$nodesize
[1] 15
> pbc_RSF$ensemble
[1] "all"
> vimp(pbc_RSF)$importance
              event.1     event.2
trt        -0.0005813 -3.286e-04
age         0.0339547  8.339e-03
sex        -0.0001609  3.283e-04
ascites     0.0049985  1.155e-02
hepato      0.0058943  1.540e-03
spiders    -0.0006503  8.148e-04
edema      -0.0040505  1.246e-02
bili        0.1707255  6.791e-02
chol        0.0028735  4.320e-03
albumin    -0.0002707  6.336e-03
copper      0.0221803  1.785e-02
alk.phos    0.0043246  5.737e-04
ast         0.0044572  2.291e-03
trig        0.0008631  8.125e-04
platelet    0.0051748  8.225e-05
protime     0.0159789  6.738e-03
stage       0.0092512  4.422e-03
> # Error rate convergence
> head(pbc_RSF$err.rate,5)
```

```
      event.1 event.2
[1,]   0.3882  0.2658
[2,]   0.3980  0.2571
[3,]   0.3793  0.2521
[4,]   0.3192  0.2381
[5,]   0.3041  0.2215
> tail(pbc_RSF$err.rate,10)
        event.1 event.2
[491,]   0.2068  0.1661
[492,]   0.2068  0.1661
[493,]   0.2068  0.1661
[494,]   0.2068  0.1662
[495,]   0.2054  0.1662
[496,]   0.2072  0.1663
[497,]   0.2054  0.1663
[498,]   0.2049  0.1665
[499,]   0.2049  0.1666
[500,]   0.2044  0.1668
> pdf("../Output/PBC_RSF_Error_Convergence.pdf",
+    height=10,width=15)
> plot(pbc_RSF,plots.one.page = TRUE)
> dev.off()
null device
          1
> # Visualizing the Random Forest Trees
> plot(get.tree(pbc_RSF,81),
+    title=paste("Survival Tree Number",81))
> plot(get.tree(pbc_RSF,265),
+    title=paste("Survival Tree Number",265))
> plot(get.tree(pbc_RSF,488),
+    title=paste("Survival Tree Number",488))
> # Variable Selection
> max.subtree(pbc_RSF)$topvars
[1] "ascites" "edema"          "protime"
> var.select(pbc_RSF, method = "md")
minimal depth variable selection ...
-----------------------------------------------------------
family              : surv-CR
var. selection      : Minimal Depth
conservativeness    : medium
x-weighting used?   : TRUE
dimension           : 17
sample size         : 276
ntree               : 500
nsplit              : 10
```

```
mtry                 : 5
nodesize             : 15
refitted forest      : FALSE
model size           : 7
depth threshold      : 4.951
PE (true OOB)        : 20.44 16.68

Top variables:
        depth vimp
bili     1.990  NA
copper   3.426  NA
ascites  4.084  NA
albumin  4.388  NA
protime  4.478  NA
edema    4.568  NA
chol     4.910  NA
-----------------------------------------------------------
> var.select(pbc_RSF, method = "vh")
> #takes lot of time
-------------------- Iteration: 1  --------------------
 selecting variables using Variable Hunting ...
 PE: 37.37     dim: 4
-------------------- Iteration: 2  --------------------
 selecting variables using Variable Hunting ...
 PE: 47.19     dim: 3
-------------------- Iteration: 3  --------------------
 selecting variables using Variable Hunting ...
 PE: 3.509     dim: 4

-------------------- Iteration: 49  --------------------
 selecting variables using Variable Hunting ...
 PE: 40.59     dim: 4
-------------------- Iteration: 50  --------------------
 selecting variables using Variable Hunting ...
 PE: 69.84     dim: 4
fitting forests to final selected variables ...
-----------------------------------------------------------
family               : surv-CR
var. selection       : Variable Hunting
conservativeness     : medium
dimension            : 17
sample size          : 276
K-fold               : 5
no. reps             : 50
nstep                : 1
```

```
ntree               : 500
nsplit              : 10
mvars               : 4
nodesize            : 2
refitted forest     : TRUE
depth ratio         : 0.0643
model size          : 3.3 +/- 1.111
PE (K-fold)         : 38.49 +/- 20.46

Top variables:
        depth rel.freq
age     1.281      30
protime 1.249      28
trt     1.258      26
hepato  1.281      26
-----------------------------------------------------------
> var.select(pbc_RSF, method = "vh.vimp", nrep = 50)
> #takes lot of time
-------------------- Iteration: 1  --------------------
 selecting variables using Variable Hunting (VIMP) ...
 PE: 53.21     dim: 2
-------------------- Iteration: 2  --------------------
 selecting variables using Variable Hunting (VIMP) ...
 PE: 46.24     dim: 2
-------------------- Iteration: 3  --------------------
 selecting variables using Variable Hunting (VIMP) ...
 PE: 41.82     dim: 3

-------------------- Iteration: 49  --------------------
 selecting variables using Variable Hunting (VIMP) ...
 PE: 40.22     dim: 2
-------------------- Iteration: 50  --------------------
 selecting variables using Variable Hunting (VIMP) ...
 PE: 37.04     dim: 2
fitting forests to final selected variables ...
-----------------------------------------------------------
family              : surv-CR
var. selection      : Variable Hunting (VIMP)
conservativeness    : medium
dimension           : 17
sample size         : 276
K-fold              : 5
no. reps            : 50
nstep               : 1
ntree               : 500
```

```
nsplit          : 10
mvars           : 4
nodesize        : 2
refitted forest : TRUE
model size      : 1.96 +/- 0.57
PE (K-fold)     : 39.64 +/- 18.92


Top variables:
        rel.freq
chol         24
hepato       22
```

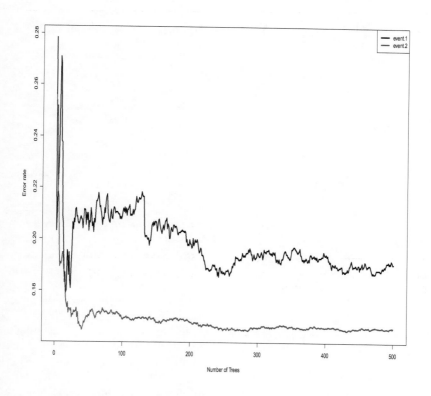

FIGURE 9.4: Random Forests Error Convergence Rate

Figure 9.4 shows that error rate convergence occurs after 300 trees. While it is not possible to extract all 500 trees of the forest, comparing Figure 9.5 with Figure 9.2, we can see that trees of the forest are more spread out. The trees of

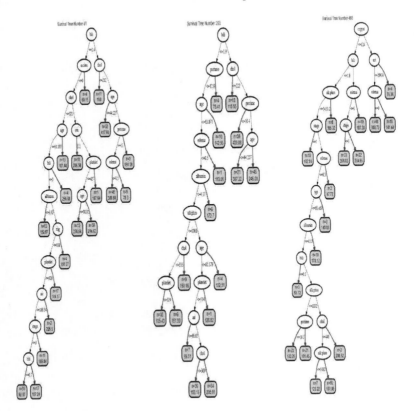

FIGURE 9.5: Random Forests Survival Trees

the random forest show more diversity than the bagging algorithm trees. Thus, we can expect the under represented variables to contribute in the fitted model. The random forest and the bagging concur on the important variables in this application.

We leave it to the reader as an exercise to complete prediction with fitted random forests and draw the comparisons with the bagging predictions. □

Chapter 4 of Tattar (2018)[112] goes into details of random forest nuances and it provides interesting diagnostics. The R package `randomForestExplainer` is useful in that direction. However, we can not use it with random survival forests. The measures explained in that package include minimum depth distribution, mean minimum depth, number of nodes in the forest for the variable, etc. These are useful and need to be extended to random survival forests. We will next move to the boosting method.

9.5   Boosting Algorithm

Classification problem has motivated several innovative solutions in the machine learning domain. Trees, neural networks, support vector machine, etc. are some of the examples which have seen remarkable innovation and adaptation toward the classification problem. The *adaptive boosting algorithm* is another example of the same journey. The boosting algorithm has since been extended to regression and survival data among a host of other solutions. The boosting algorithm works differently from the earlier two ensemble methods.

In the context of the classification problem, the adaptive boosting method works as follows. Instead of trees, we use simple stumps where the classification is carried out using a single variable. It is obvious that in most practical classification problems, the stumps would be poor classifiers, and they are also aptly called as *weak learners*. Thus, if a categorical variable has k levels, we can have $k - 1$ different stumps. Given a weak learner, the adaptive boosting algorithm gives zero weight to the correctly classified observations and distributes more weight to the misclassified observations. Moving on to other stumps, the boosting algorithm would each time update the weights for the misclassified observations and gradually convert the weak learners into an overall strong ensemble. The work of Freund and Schapire (1996)[45] has been path breaking and it has been recipient of well recognized awards too. Unlike the other two ensembles dealt with earlier, we have a sequential approach to the boosting algorithm.

The adaptive boosting algorithm works fine only for the classification problem. The approach needs adaptation and we will introduce the general gradient boosting algorithm in the subsection that follow.

9.5.1   Gradient Boosting Algorithm

The gradient boosting algorithm extends the ideas of the adaptive boosting algorithm and by making use of loss functions, it can be applied to a variety of problems. Thus, we can use the *gradient boosting method*, popularly known as *GBM*, not only for the regression problem but for the classification problem too. The generalization to the regression was accomplished in Friedman (2001)[46]. Ridgeway (2007)[93] shows how the loss functions idea helps in applying GBM to the following classes of problems: the (Gaussian) regression, adaptive boosting, binary classification, Laplace, quantile regression, survival data, and Poisson regression.

Chapter 5 of Tattar (2018)[112] illustrates the bare bones adaptive boosting method for the classification problem and then proceeds to demonstrate the GBM under the squared-error loss function. We will state the squared-error loss form of GBM here to outline its working.

Let **y** denote the output vector of n observations and let $\boldsymbol{X}$ be the regression matrix of dimension $n \times (p+1)$. We will run the GBM for B iterations, or using B trees. Let the depth of the tree be d and the shrinkage factor be ϵ. The GBM algorithm under the squared-error loss can be then succinctly expressed in the following, see Algorithm 17.2 of Efron and Hastie (2016)[41] or Section 'Gradient boosting', Chapter 5 of Tattar (2018)[112].

ALGORITHM 9.5.1. **Gradient Boosting Method Under Squared-Error Loss Function**.

1. Initialize residuals $\mathbf{r} = \mathbf{y}$ and set the gradient boosting prediction as $\hat{G}_0 = 0$.

2. For $b = 1, 2, \ldots, B$:

    (a) Fit a regression tree of depth d for the data $(\mathbf{r}, \mathbf{X})$.

    (b) Obtain the predicted values as $\hat{G}_b$.

    (c) Update the boosting prediction by $\hat{G}_b = \hat{G}_{b-1} + \epsilon \hat{G}_b$.

    (d) Update the residuals by $\mathbf{r} = \mathbf{r} - \hat{G}_b$.

3. Return the sequence of functions $\hat{G}_b, b = 1, 2, \ldots, B$.

The GBM under the squared-error loss function can be easily setup, reproduced from Section 'Gradient boosting', Chapter 5 of Tattar (2018)[112], as follows:

```
> # Gradient Boosting Using the Squared-error Loss Function
> GB_SqEL <- function(y,X,depth,iter,shrinkage){
+    curr_res <- y
+    GB_Hat <- data.frame(matrix(0,nrow=length(y),ncol=iter))
+    fit <- y*0
+    for(i in 1:iter){
+       tdf <- cbind(curr_res,X)
+       tpart <- rpart(curr_res~.,data=tdf,maxdepth=depth)
+       gb_tilda <- predict(tpart)
+       gb_hat <- shrinkage*gb_tilda
+       fit <- fit+gb_hat
+       curr_res <- curr_res-gb_hat
+       GB_Hat[,i] <- fit
+    }
+    return(list(GB_Hat = GB_Hat))
+ }
```

The left panel of Figure 9.6 displays the sine wave. Now, if one uses the transformation, the linear model suffices. However, the question here is whether using simple trees and without having a preview of the scatterplot, is it possible to approximate the relationship between Y and x's under the GBM? We can see that the residuals keep on decreasing in the right panel of the same figure. Thus, we see that GBM works.

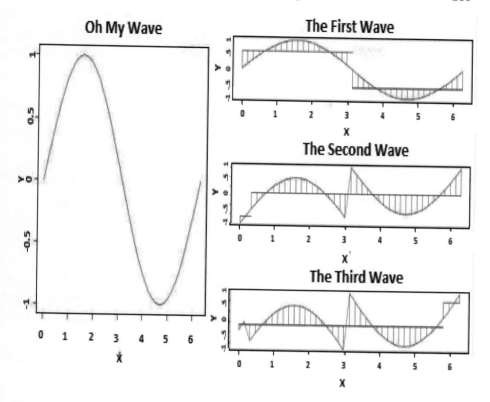

FIGURE 9.6: Approximating Sine Wave with Boosting

Now, let $\mathbf{x}$ denote the vector of features, Y the output variable of interest, $f(\mathbf{x})$ the function mapping the feature vector to the regressand, and $\Psi(y, f(\mathbf{x}))$ the loss function. We continue to assume that there are n observations. The general GBM, algorithm stated from Ridgeway (2007)[93], can be then stated in the following form.

ALGORITHM 9.5.2. **Generic Gradient Boosting Algorithm.**
1. Initialize $\hat{f}(\mathbf{x})$, say with $\hat{f}(\mathbf{x}) = \arg\min_\rho \sum_{i=1}^n \Psi(y_i, \rho)$.
2. For $b = 1, \ldots, B$, repeat:

   (a) Obtain the negative gradient as follows:

   $$z_i = -\frac{\partial}{\partial f(\mathbf{x}_i)} \Psi(y_i, f(\mathbf{x}_i)) \Big|_{f(\mathbf{x}_i) = \hat{f}(\mathbf{x}_i)} \tag{9.1}$$

   (b) Fit a regression model $g(\mathbf{x})$ and predict z_i from the covariates $\mathbf{x}_i$.
   (c) Calculate the gradient descent step by

   $$\rho = \arg\min_\rho \sum_{i=1}^n \Psi\left(y_i, \hat{f}(\mathbf{x}_i) + \rho g(\mathbf{x}_i)\right). \tag{9.2}$$

3. Update $\hat{f}(\mathbf{x})$ by

$$\hat{f}(\mathbf{x}) \leftarrow \hat{f}(\mathbf{x}) + \rho g(\mathbf{x}). \tag{9.3}$$

We will next illustrate application of the GBM with the Cox PH model.

9.5.2   Boosting the Cox PH Model

Ridgeway (2007)[93] extended the GBM to boosting the Cox PH model and provided an implementation of the same in the **gbm** R package. Section 4.6 of Ridgeway provides the details related to the deviance, gradient, initialization, Hessian matrix, update rule, etc. The boosting algorithm related to the Cox proportional hazards regression model is implemented in **gbm** package created by Ridgeway. We need to merely specify the distribution option as "coxph" alongwith the survival object and experiment with the number of trees. We work with the PBC data as earlier.

Example 51 *Applying the Gradient Boosting Modeling to the PBC Data. Using the* gbm *function from the same named package, we are able to execute the GBM for the survival data. We experiment with the default settings and then change the number of trees to 200. The cross-validation alternative is also used and we evaluate a* gbm *with it.*

```
> # GBM for the Cox PH Model
> pbc_gbm <- gbm(Surv(time,status==2)~.,dist="coxph",
+    data=pbc[,-1])
> pdf("../Output/Relative_Importance.pdf",height=10,width=25)
> par(mfrow=c(1,3))
> summary(pbc_gbm)
                var rel.inf
bili           bili 29.6569
protime     protime  9.5722
albumin     albumin  9.4121
copper       copper  8.9428

hepato       hepato  0.0000
> title("The Generic Relative Influence Plot")
> length(pbc_gbm$trees)
[1] 100
> # Increasing the number of trees
> pbc_gbm2 <- gbm(Surv(time,status==2)~.,dist="coxph",
+                 n.trees=200,data=pbc[,-1])
> summary(pbc_gbm2)
                var rel.inf
bili           bili 18.6217
protime     protime 14.3011
```

```
copper        copper 12.6996
platelet platelet 10.4975

spiders     spiders  0.0000
> title("Relative Influence With More Trees")
> length(pbc_gbm2$trees)
[1] 200
> # GBM with Cross-validations
> pbc_gbm3 <- gbm(Surv(time,status==2)~.,dist="coxph",
+                 cv.folds = 10, data=pbc[,-1])
> summary(pbc_gbm3)
             var rel.inf
bili        bili  27.690
copper    copper  14.208
protime  protime  11.051
age          age   8.624

spiders     spiders  0.000
> title("Relative Influence with Cross-Validataions")
> length(pbc_gbm3$trees)
[1] 100
> dev.off()
```

Figure 9.7 gives the relative influence of the variables in each ensemble of pbc_gbm, pbc_gbm2, *and* pbc_gbm3. *Note that we did not use the* plot *function at all. Simply running the* summary *function gets the relative influence plots on the required device.*

□

Predictions are possible only at the link level with the gbm objects. The original response scale, of the time to event, predictions are not possible with the current package.

9.6 Exercises

Exercise 9.1 *Change the split rule in Example 9.2 to* logrankscore *and create the bagging ensemble. Generate the bagging convergence rate and compare it with Figure 9.3. How many of the top variables in* pbc_bagging *remain with the logrank score split?* □

Exercise 9.2 *Compare the error convergence rate of* pbc_bagging *in Figure 9.3 with the random forest* pbc_RSF *error rate of Figure 9.4.* □

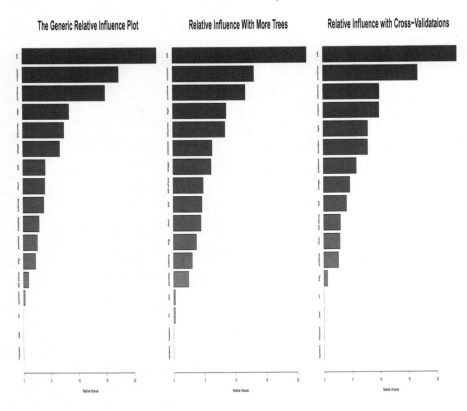

FIGURE 9.7: Relative Influence of Variables in GBM

Exercise 9.3 *Create bagging, random forest, and gradient boosting ensembles for Netherlands Cancer Institute Seventy Gene Signature, see Subsection 1.2.6. Compare the error convergence plots.*   □

9.7   Más Lejos Temas

Ensemble learning is a powerful technique wherein we do not have to worry about selecting the "best" model. By combining models that are minimally accurate in a certain sense, the ensemble provides a more robust and accurate model. A random forest based on likelihood ratio test statistic split is available in the R package `icRSF`.

A widely unstated assumption of ensemble learning is that the contributing models are independent of each other. It is difficult to ensure independence of models even if they are drawn on completely different samples. In case of ensemble learning techniques such as bagging and random forest, the bootstrap sample is expected to consist of $0.63 \times n$ unique observations on an average, and between two bootstrap samples around $0.4 \times n$ common observations. How can two trees be independent of each other then? While it is a nearly impossible task to test for the independence of trees, it is enough that two trees must not predict very closely. This is because of the powerful adage that if two people agree with each other all the time, one of them is not required. In the context of classification and regression trees, a few methods based on statistical measures of agreement and deviance tests, it is possible to carry out diagnostics on independence of the trees. However, we are not aware of any such methods for survival forests or bagging survival trees. The enthusiastic reader can consult Kuncheva (2014)[67] for more details on diagnostic tests related to ensembles. Tattar (2018)[112] also has a chapter dedicated to ensemble methods for survival data.

Further details of boosting techniques can be found in Schapire and Freund (2012)[100]. Multi-state models can also be enriched by this ensemble technique and an R implementation of the same is available in the `gamboostMSM` package.

Chapter 10

Neural Network Survival Analysis

10.1 Introduction

Neural network forms one of the most important techniques in the field of Machine Learning. The motivation for neural networks, abbreviated as NN, stems from the working of neurons in the brain and the network formed by them. The brain encompasses true neural networks, while machine learning techniques are artificial neural networks, we will mean the latter whenever we say neural network.

The first version of the neural network model is the simplistic perceptron model. In a simple network, the input layer is made of covariates as in the neurons, the arc weights connect the input layer to the output layer, and the output layer consists of a single neuron which is the output variable. An activation function accumulates the weighted inputs. The perceptron model can learn linear patterns but it is not able to handle nonlinear patterns. We will see later in the chapter that multiple layers are important for a NN to handle nonlinear data. Construction of a neural network relies heavily on the notion of *activation functions*. Use of the sigmoid activation function has lead to considerable research leading to improvisation in application of NN to classification problems. Considerable research has gone into improvisation of ML towards use of NN in addressing classification problems. Technically, it is possible to have multiple outputs in NN with the same architecture. In survival analysis, this flexibility translates into one output for the time to event and a second output for the event indicator. However, most software implementations do not support such a task.

Neural networks have been applied to survival data from the 1990's. Ripley (1993)[94], Cheng and Titterington (1996)[23], and Schwarzer et al. (2000)[102] contain some adaptations of NN to analysis of survival data. Ripley and Ripley (2007)[95] provide a comprehensive review of the NN methods, and it is a contributory paper in Dybowski and Gant (2007)[38] which is a compilation of different applications of NN in clinical studies. Ripley and Ripley (2007)[95] lament the misuse of classification NN for survival data. They note that a number of practitioners bin the lifetimes in a few classes of age intervals and then apply the usual NN which implements the multi-classification problem.

DOI: 10.1201/9781003306979-10

In the next section, we will discuss the general architecture of an NN. Activation functions will be discussed in Section 10.3. The basic NN in the form of the perceptron model will be taken up in Section 10.4. The multilayer NN will be setup in Section 10.5. We will conclude the chapter with an indication of how to use the NN for survival data in Section 10.6.

10.2   Neural Network Architecture

The setup of neurons in the brain as seen in Figure 10.1, reproduced from Figure 2 of Haykin (2009)[52] has inspired the architecture of neural networks. The figure is an illustration of the shape of pyramidal cell which is one of the common types of cortical neurons. The cortical neuron receives the input through dendritic spines. The cell receives more than 1e5 synaptic contacts and it can further project onto thousands of target cells. Most of the neurons pass on their output through a series of brief voltage pulses/action potentials/spikes. The spikes originate closer to the cell body of neurons and then propagate across the individual neurons at almost constant velocity and amplitude. The use of action potentials as communication between the neurons is based on the physics of axon which is very long and thin and characterized by high electrical resistance and large capacitance. This description is adapted from Haykin (2009)[52].

The success of this machine learning technique has multiple reasons and has many useful practical properties:

- **Nonlinearity**: NN can easily handle linear relationships and also in constructing and analyzing nonlinear models.

- **Input-Output Relationship**: It is an example of supervised learning and can be effectively used as a regression tool.

- **Adaptivity**: NN adapts, imitates the biological neurons, and effects changes in synaptic weight changes as the environment changes.

- **Evidential Response**: It helps in identifying useful covariates and also gives the probability associated with the class.

- **Contextual Information**: In a certain sense, it gives more weight to the evidence in neighborhood.

- **Fault Tolerance**: Imitates real neuron in the sense that an NN works even if some of the covariates are lost completely.

- **VLSI Implementation**: Since an NN imitates the biological neuron functioning, it is useful in simulation of high electrical resistance and very large capacitance.

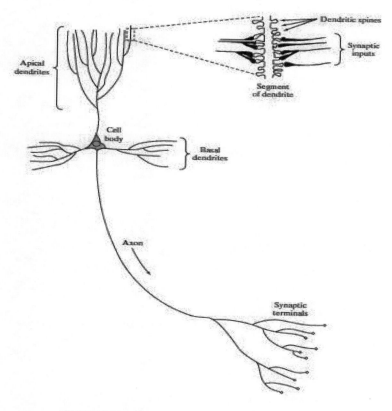

FIGURE 10.1: Biological Neural Network

The core structure of an artificial neuron, neural network architecture, is given in Figure 10.2. A neuron is constructed as an essential information-processing unit in the working of a neural network. The three most important blocks of a neuron are described next:

- A set of synapses comes along with its own weight. That is, an input/signal x_i is available at the synapse which is connected with the neuron and the corresponding synaptic weight is β_i. We also have a synapse with an unit value input to the neuron which comes with a weight of β_0.

- All the effective signals received by the neuron are added up and the operation is the linear combiner Σ.

- An activation function ϕ for limiting the amplitude of the output of a neuron.

Let us consider a single observation with output value y, which corresponds to the realized value of the random variable Y, and p covariate values of $x_1, x_2, \ldots, x_p$.

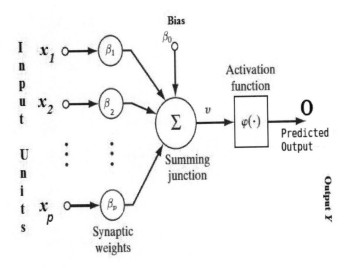

FIGURE 10.2: The Structure of Neural Network

Mathematically, the input received by a neuron is summarized by

$$v = \sum_{i=1}^{p} \beta_j x_j.$$

The output predicted by the neural network, for given weights $(\beta_0, \beta_1, \ldots, \beta_p)$, is then

$$O = \phi(v + \beta_0).$$

In statistical terms, O is the fitted value of the neural network. The effect of using the bias term is in carrying out an *affine transformation* to the input v. Note that the term bias used in this context is entirely different from the form used in statistical theory. The input x_0 associated with the bias is defined as 1.

The core idea is then to tune in the weights $(\beta_0, \beta_1, \ldots, \beta_p)$ such that $y - O$ is as less as possible. Before we pursue it, we need to have detailed discussion of the activation functions ϕ.

10.3   Activation Function

Activation functions are key to the success of the neural networks. Input v to a neuron passes out of it following action of the activation function. Depending on the type of problem one is trying to solve, the activation function will vary.

The main types of activation function are the following:

1. Linear Activation Function

2. Hard Limit Activation Function

3. Sigmoid Activation Function

4. Hyperbolic Tangent Function

5. Rectified Linear Unit Activation Function

6. Leaky Rectified Linear Unit Activation Function

The *linear activation function* is given by

$$\phi(v) = v, \forall \quad v \in R^1.$$

This function takes values on the real line, and it is especially useful when the output variable is a numeric variable. Also, in multiple layer neural network, see Section 10.5, it is recommended to use linear activation function up to the output layer. Linear activation function is used whenever Y is a numeric variable and it is preferred choice for all the $l - 1$ layers of neurons when there are l layers.

The *hard limit activation* is parametrized by a constant value c and is defined by

$$\phi(v, c) = \begin{cases} 1, & \text{if } v > c, \\ 0, & \text{if } v \leq c. \end{cases}$$

The hard-limit function is seen to be a jump function and it is useful in threshold classification, or the general classification problem. Any reader with exposition to empirical distribution function would be familiar with jump function.

The *sigmoid activation function* is given by

$$\phi(v) = \frac{1}{1 + e^{-v}}.$$

It can be seen that as $v \downarrow -\infty$, $\phi(v) \downarrow 0$, and as $v \uparrow \infty$, $\phi(v) \uparrow 1$. Thus, the output of the sigmoid activation function is in the interval $[0, 1]$ making it convenient to employ the function in classification problems. The sigmoid activation function is popular for another reason too. In the backpropogation algorithm step, we need the derivative value of the activation function which for the sigmoid function is easily obtained as $\phi'(v) = \phi(v)(1 - \phi(v))$, the proof is given in Section 10.5, and thus it is easier to compute. When the sigmoid activation function is used for classification problem, the prediction threshold is 0.5, that is, when the output value $\phi(v)$ of the activation function is greater than 0.5, the class is predicted as 1, and 0 otherwise.

The *hyperbolic tangent activation function* is given by

$$\phi(v) = \frac{e^v - e^v}{e^v + e^v}.$$

Now, as $v \downarrow -\infty$, $\phi(v) \downarrow -1$, and as $v \uparrow \infty$, $\phi(v) \uparrow 1$. This activation function is commonly used for classification problem, and here the threshold will be 0. That is, when $\phi(v) > 0$, the class is predicted as 1, and 0 otherwise.

In theory, each neuron can have a different activation function even when it belongs to the same layer. However, most software implementations force a single kind of activation function for the neurons belonging to the same layer. As such, we do not have mathematical proof of which activation function to use at which layer. The type of regression problem dictates the kind of activation function for the output layer. In our observation, it is faster, if not convenient, to use linear activation function in the hidden layers. Using multiple layers with linear activation function will not lead to improvements.

For image classification and further applications to autonomous car drive, the sigmoid functions have been seen not to give satisfactory results. A new function, useful in *deep learning*, has become popular these days. *Rectified linear unit* activation function has been found to be very useful in modern applications. It is a variant of the linear unit function with the difference being that negative values are reported at zero:

$$\phi(v) = \max(0, v).$$

While the negative values are forced to be zero, the leaky rectified linear unit activation function tries to diminish the negative values by reducing them by a certain factor a, that is,

$$\phi(v, a) = \begin{cases} v, & \text{if } v > 0, \\ a \times v, & \text{if } v \leq 0. \end{cases}$$

Experimenting with different activation functions is the only way to obtain sufficient accuracy.

The R codes for these six activation functions are given in the code block below. The values are computed over a grid and then the function output values are plotted against these values. The output of the following code block is Figure 10.3.

```
> # Activation Function
> pdf("../Output/Activation_Functions.pdf")
> par(mfrow=c(3,2))
> # Linear Activation Function
> Lin_Af <- function(x) x
> x <- seq(-5,5,0.1)
> plot(x,Lin_Af(x),"l",xlab="x",ylab="Linear Activation")
> title("Linear Activation Function")
```

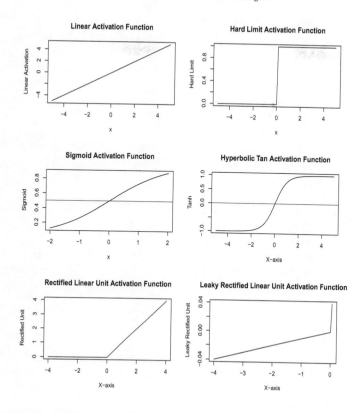

FIGURE 10.3: Activation Functions for Neural Network

```
>
> # Hardlim Activation Function
> x <- seq(-5,5,0.1)
> HL_Af <- function(x) ifelse(x>0,1,0)
> plot(x,HL_Af(x),"l",xlab="x",ylab="Hard Limit")
> title("Hard Limit Activation Function")
>
> # Sigmoid Activation Function
> Sig_Af <- function(x) 1/(1+exp(-x))
> x <- seq(-2,2,0.01)
> plot(x,Sig_Af(x),"l",xlab="x",ylab="Sigmoid")
> abline(h=0.5,col="red")
> title("Sigmoid Activation Function")
>
> # Hyperbolic Tan Activation Function
> HyperTan_Af <- function(x) tanh(x)
```

```
> x <- seq(-5,5,0.1)
> plot(x,HyperTan_Af(x),"l",xlab="X-axis",ylab="Tanh")
> abline(h=0,col="red")
> title("Hyperbolic Tan Activation Function")
>
> # Rectified linear unit
> ReLu_Af <- function(x) ifelse(x<0,0,x)
> x <- seq(-4,4,0.1)
> plot(x,ReLu_Af(x),"l",xlab="X-axis",ylab="Rectified Unit")
> title("Rectified Linear Unit Activation Function")
>
> # Leaky Rectified linear unit
> Leaky_ReLu_Af <- function(x) ifelse(x<0,0.01*x,x)
> x <- seq(-4,0.04,0.01)
> plot(x,Leaky_ReLu_Af(x),"l",xlab="X-axis",
+    ylab="Leaky Rectified Unit")
> title("Leaky Rectified Linear Unit Activation Function")
>
> dev.off()
null device
          1
```

Activation functions are required from the first (hidden) layer until output layer. The non-continuous activation functions are recent developments and they are more useful with new advances such as deep learning. We next look at the simple form of an NN.

10.4   Perceptron Model

The discussion here borrows heavily from Rojas (1996)[98]. This 1996 classic also provides comprehensive mathematical details associated with NN. We will begin with a definition of the learning algorithm.

DEFINITION 10.4.1. A *learning algorithm* is an adaptive method by which a network of computing units self-organizes to implement the desired behavior.

It is important to understand the definition which says that the algorithm is an adaptive method, and that it will gradually improve and accomplish the goal. Typically, the output of learning algorithm is accompanied by a learning rate too. The complexity parameter in the context of decision/survival trees prevents the algorithm from splitting further if the improvisation does not cross a certain threshold. In the context of neural networks, the error correction for a given sample is restricted to a fraction so that we do not over-shoot a reasonable threshold for the error at every instance of error correction.

Learning of the algorithm happens through some examples of desired input-output mapping. We continue with the classification problem. A correction step is executed iteratively until the network learns to produce the desired response. The notations are briefly stated again:

- $(x_1, x_2, \ldots, x_p)$: input to the perceptron is the covariate vector.

- $\beta_1, \beta_2, \ldots, \beta_p$: weights of the perceptron, generally used notation in NN literature is $w_1, w_2, \ldots, w_p$.

- β_0: bias, in NN notations it is w_0.

- $(x_1, x_2, \ldots, x_p, 1)$: extended covariate vector.

- $\boldsymbol{\beta} = (\beta_1, \beta_2, \ldots, \beta_p, \beta_0)$: extended weight vector.

The perceptron algorithm with input weights $(\beta_1, \beta_2, \ldots, \beta_p)$ and the covariate vector $(x_1, x_2, \ldots, x_p)$ checks whether

$$(\beta_1, \beta_2, \ldots, \beta_p)'(x_1, x_2, \ldots, x_p) \geq \beta_0.$$

With the extended weight vector $\boldsymbol{\beta} = (\beta_1, \beta_2, \ldots, \beta_p, \beta_0)$ and extended covariate vector $(x_1, x_2, \ldots, x_p, 1)$, the check is

$$(\beta_1, \beta_2, \ldots, \beta_p, \beta_0)'(x_1, x_2, \ldots, x_p, 1) \geq 0.$$

The above check is carried out for all the observations. The perceptron model uses an algorithm, described in Algorithm 10.4.1, in which the previous check is used to update the coefficient vector. The NN literature consistently calls out the regression coefficients as weights and we will use the NN terminology here. The simple perceptron model will be able to learn the patterns where the positive and negative cases are linearly separable. Thus, if we consider the classification problem of Figure 8.2, the perceptron model will not be able to classify with reasonable accuracy. Mathematically, we will define linear separability in the following.

DEFINITION 10.4.2. **Linearly Separable**. Two sets of points A and B in a p-dimensional space are called *linearly separable* if there exist $p + 1$ real numbers $\beta_1, \beta_2, \ldots, \beta_p, \beta_0$ such that every point $(x_1, x_2, \ldots, x_p) \in A$ satisfies $\sum_{j=1}^{p} \beta_j x_j \geq \beta_0$ and every point $(x_1, x_2, \ldots, x_p) \in B$ satisfies $\sum_{j=1}^{p} \beta_j x_j < \beta_0$.

Note that the definition says nothing about classification problem or learning algorithm. It simply spells out, by a criterion, a rule to declare two collections of points as linearly separable. For our classification problem, it would be ideal to have the classes as linearly separable which would be separated if we can find the $p + 1$ real numbers $\beta_1, \beta_2, \ldots, \beta_p, \beta_0$. A slightly lesser restrictive condition of separability is given next.

DEFINITION 10.4.3. **Absolutely Linearly Separable**. Two sets of points A and B in a p-dimensional space are called *absolutely linearly separable* if there exist $p + 1$ real numbers $\beta_1, \beta_2, \ldots, \beta_p, \beta_0$ such that every point $(x_1, x_2, \ldots, x_p) \in A$ satisfies $\sum_{j=1}^{p} \beta_j x_j > \beta_0$ and every point $(x_1, x_2, \ldots, x_p) \in B$ satisfies $\sum_{j=1}^{p} \beta_j x_j < \beta_0$.

We will now look at the perceptron learning algorithm which will solve the classification problem in case of linearly separable points.

10.4.1 Perceptron Learning Algorithm

Let P and N stand for two finite set of points in $\mathbf{R}^p$. Connecting with the previous definitions, we might take $A = P$ and $B = N$. To linearly separate the two sets, we seek a weight vector so that the points in P belong to the vector's positive half-space, and the points in N to the negative half-space.

The *error* of a perceptron with weight-vector $\boldsymbol{\beta}$ is the number of incorrectly classified points. Here, the number in parenthesis ( ) indicates the iteration number. Also, the operator $:=$ in a $:=$ b means that the value of b is now assigned to a. The perceptron learning algorithm is given in the following:

ALGORITHM 10.4.1. **Perceptron Learning Algorithm**

1. Generate the weight vector $\boldsymbol{\beta}^{(0)}$ randomly and set $t := 0$.

2. A vector $\mathbf{x} \in P \cup N$ is selected randomly, and

   (a) If $\mathbf{x} \in P$ and $\boldsymbol{\beta}^{(t)}\mathbf{x} > 0$, go to the beginning of this step.
   (b) If $\mathbf{x} \in P$ and $\boldsymbol{\beta}^{(t)}\mathbf{x} \leq 0$, go to Step 3.
   (c) If $\mathbf{x} \in N$ and $\boldsymbol{\beta}^{(t)}\mathbf{x} < 0$, go to the beginning of this step.
   (d) If $\mathbf{x} \in N$ and $\boldsymbol{\beta}^{(t)}\mathbf{x} \geq 0$, go to Step 4.

3. Set
$$\boldsymbol{\beta}^{(t+1)} = \boldsymbol{\beta}^{(t)} + \mathbf{x}, \text{ and } t := t + 1,$$
and go to Step 2.

4. Set
$$\boldsymbol{\beta}^{(t+1)} = \boldsymbol{\beta}^{(t)} - \mathbf{x}, \text{ and } t := t + 1,$$
and go to Step 2.

Why should this algorithm guide the user toward separation of the P's from the N's? When the hyperplane that separates the two sets of points is misclassifying a positive observation, the updated weight vector on being added with the observation will not misclassify it. Similar logic applies for the negative points on being misclassified. Because the observations are linearly separable, we will eventually have the weight vector that will not misclassify a point any further. The perceptron convergence theorem guarantees that if the two sets P and N are linearly separable, vector $\boldsymbol{\beta}$ is updated only a finite number of times.

Three simplified assumptions have been found useful in applications:

- The sets P and N are merged in a single set $P' = P \cup N^-$, where the set N^- contains the negated elements of N, here negated means the sign of the elements are reversed.

- The vectors in P' are normalized.

- The weight vector is also normalized.

We will illustrate the perceptron model for the simple OR logic. With two binary variables, the OR logic returns the output as 1 if at least one of the variables has a value of 1. If X_1 and X_2 are binary inputs and Y is the output, Y takes the value 1 if any of the inputs is having value 1. Let's check out the algorithm in action next.

The input variables x1, x2, and the intercept variables x0 are first created and then combined into a regression matrix X. The OR logic output y is defined and the weights are initialized in w.

```
> # Learning the OR rule
> x0 <- c(1,1,1,1)
> x1 <- c(1,1,0,0)
> x2 <- c(1,0,1,0)
> y <- c(1,1,1,0)
> X <- cbind(x0,x1,x2)
> # Initialize the weights
> w <- c(1,1,1)
> alpha <- 0.2
>
> X%*%w
     [,1]
[1,]    3
[2,]    2
[3,]    2
[4,]    1
> Predictions <- sign(X%*%w)
> Predictions
     [,1]
[1,]    1
[2,]    1
[3,]    1
[4,]    1
```

We can see that the predictions are all 1, and hence the last combination of the OR logic is misclassified. We first define the `Update_Weights` function and then keep on updating the output and the weight vector until all the observations are correctly predicted. Here, `alpha` is the learning rate.

```
> Update_Weights <- function(y,X,w,alpha){
+    dw <- t(X)%*%(alpha*(y-Sig_Af(X%*%w)))
+    uw <- w+dw
+    return(uw)
+ }
> delta <- sum(y-Sig_Af(X%*%w))^2
> while(delta>1e-4){
+    w <- Update_Weights(y,X,w,alpha)
+    delta <- sum(y-Sig_Af(X%*%w))^2
+    print(delta)
+ }
[1] 0.1717
[1] 0.1488
[1] 0.1292
[1] 0.1125
[1] 0.09831

[1] 0.0001011
[1] 0.0001007
[1] 0.0001003
[1] 9.993e-05
> X%*%w
        [,1]
[1,] 10.783
[2,]  3.911
[3,]  3.911
[4,] -2.960
> Predictions <- Sig_Af(X%*%w)
> round(Predictions,1)
      [,1]
[1,]    1
[2,]    1
[3,]    1
[4,]    0
```

We thus see that the perceptron model can easily learn the OR logic. The perceptron model cannot handle nonlinear data and we need multi-layer neural networks for that.

10.5   Multi-layer Neural Network

The strength of the machine learning techniques has been the ability to handle nonlinear data. Importantly, the nonlinear machine learning techniques differ drastically from the conventional nonlinear statistical techniques, see for instance Seber and Wild (1989)[103], wherein the form of the nonlinear relationship is clearly known. Where the domain knowledge eludes creation of the functions which model the causal nonlinear relationships, generic nonlinear machine learning techniques are useful. Without knowing any semblance of relationship between the input variables and the output variables, we would like to predict the output using algorithms of specific formats. It is well-known that given enough number of data points, the multi-layer neural networks are very good universal approximators. Let us begin with a brief peek into a multi-layer NN. In a multi-layer NN, we have layers of neurons between the input variables layer and the output variables layer. Each layer will consist of a specified number of neurons. Typically, each layer will be acted upon by the same class of activation functions. Further, the number of neurons in each layer can be explicitly specified. Refer Haykin (2009)[52] or Matlab document available at `http://www.image.ece.ntua.gr/courses_static/nn/matlab/nnet.pdf` for matrix representation of neural networks. We will next define the architecture of an NN.

DEFINITION 10.5.1. A *neural network architecture* is a tuple (I, N, O, E) consisting of a set I of input variables, a set N of hidden layer neurons, a set O of output variables, and a set E of weighted directed edges. A directed edge is a tuple (u, v, w) where $u \in I \cup N, v \in N \cup O$, and $w \in \mathbf{R}$.

The ensuing development assumes sigmoid activation function throughout the network. That is, the activation functions for all neurons in the hidden layer N are sigmoid functions. Further, the input vector $\mathbf{X}$ is a p-dimensional vector and the output vector $\mathbf{Y}$ is q-dimensional.

Consider the sigmoid activation function:

$$\phi_c(x) = \frac{1}{1 + e^{-cx}},$$

where the constant c can be selected arbitrarily, and its reciprocal is called the *temperature parameter* in neural networks. We take $c = 1$ in the rest of the chapter.

The derivative of the sigmoid function w.r.t. x is

$$
\begin{aligned}
\frac{d}{dx}\phi(x) &= \frac{(-1)e^{-x}(-1)}{(1 + e^{-x})^2} \\
&= \frac{e^{-x}}{(1 + e^{-x})^2} \\
&= \phi(x)(1 - \phi(x)).
\end{aligned}
$$

Given weights $\beta_1, \ldots, \beta_p, \beta_0$, the sigmoidal function for an input vector $\mathbf{x}$ calculates:

$$\phi(\mathbf{x}, \boldsymbol{\beta}, \beta_0) = \frac{1}{1 + \exp\left(\sum_{j=1}^p \beta_j x_j + \beta_0\right)}.$$

Loosely speaking, the objective is to determine the weights $\boldsymbol{\beta}$ and β_0 such that difference between predicted output of the network in the O's and the actual Y's is a minimum. In this direction, the weights are initialized randomly. Using the initial weights, the aggregated input to a neuron is computed. Next, the value for the input is translated into the output of the neuron through the activation function, sigmoid activation function here. Simultaneously, the derivative value of the activation function is calculated and stored. This action is repeated for each and every neuron across all the layers till we obtain the predicted output of the neural network in $\mathbf{O}$. The action till this stage is called as the *forward pass*. The predicted output value $\mathbf{O}$ is then compared with the actual value $\mathbf{Y}$ and the error is calculated. Using an error loss function, the gradient of the error is passed backward layer-by-layer from the output layer to the inputs. The weights are then suitably modified and the forward-backward pass steps are repeated again and again until we reach a stability point, a point after which the overall error does not reduce any further. This is the bigger view of how the neural networks work. We explain them more formally in the following development.

A feed-forward neural network (NN) is a computational graph whose nodes are computing units and whose directed edges transmit numerical information from node to node. Each computational unit is capable of evaluating its impact through a single activation function. The learning problem consists of finding the optimal combination of weights so that the network learns the patterns.

Consider an NN with p input covariables and q output variables. Generally, we have $q = 1$ for regression problem, and q is the number of labels for classification problem. Suppose that we have a training set $\{(\mathbf{x}_1, \mathbf{y}_1), (\mathbf{x}_2, \mathbf{y}_2), \ldots, (\mathbf{x}_n, \mathbf{y}_n)\}$, where $\mathbf{x}_i, i = 1, \ldots, n$ is a p-dimensional vector, and $\mathbf{y}_i, i = 1, \ldots, n$ is a q-dimensional vector. We assume that the activation function at each node of the network is continuous and differentiable function. Initially, the weights are selected at random. With the input vector $\mathbf{x}_i$, the NN will give $\mathbf{o}_i, i = 1, \ldots, n$, as the predicted output. We seek to minimize the NN error:

$$E = \frac{1}{2} \sum_{i=1}^{n} \|\mathbf{o}_i - \mathbf{y}_i\|^2 ,$$

where $\|\mathbf{u}\|^2 = \sum_{j=1}^{q} u_j^2 / q$.

The first step of the minimization process consists of extending the network so that the network automatically calculates the error. Note that each output node is connected to a node that evaluates the function $\frac{1}{2} (o_{ij} - y_{ij})^2$, $i = 1, \ldots, n, j = 1, \ldots, q$, see Figure 10.4. The output of the additional q nodes is collected and added up to give the error E_i. The same network is built for each observation, and the overall error is $E = \sum_{i=1}^{n} E_i$. That is, the overall error is the quadratic sum:

$$E = \sum_{i=1}^{n} E_i = \frac{1}{2q} \sum_{i=1}^{n} \sum_{j=1}^{q} (o_{ij} - y_{ij})^2 .$$

EXTENDED NN

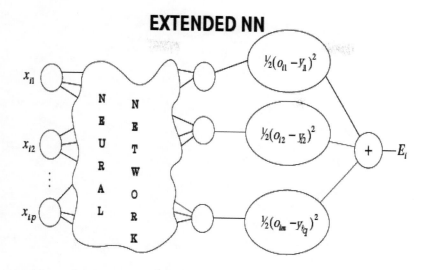

FIGURE 10.4: Neural Network with Error

The weights in the network are the only parameters that can be modified to minimize the quadratic error E. Thus E is minimized by using an iterative process, gradient descent, based on (the gradient):

$$\nabla E = \left(\frac{\delta E}{\delta \beta_1}, \frac{\delta E}{\delta \beta_2}, \ldots, \frac{\delta E}{\delta \beta_l} \right),$$

and each weight is updated by using the increment:

$$\Delta \beta_i = -\nu \frac{\delta E}{\delta \beta_i}, i = 1, \ldots, l.$$

Note that we use l as the number of weights and not p since in a multiple NN, the number of weights is also decided by the number of neurons in the hidden layer.

The NN is equivalent to a complex chain of function compositions. Hence, the chain rule of differential calculus plays a major role in finding the gradient of the error function.

DEFINITION 10.5.2. An NN node is arranged in left and right sides, with the right side function enabling the forward pass and the left side for the backward pass. This representation, see Figure 10.5, is called a *B-diagram*.

The right side computes the activation function associated with the node while the left side computes the derivative of the activation function for the same input. The computations of function f and its derivative f' as displayed in Figure 10.5. This can be further detailed as follows.

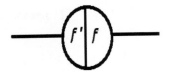

FIGURE 10.5: B Diagram of a Node

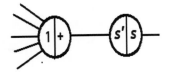

FIGURE 10.6: Activation Function—Integration and Differentiation

First, the B-diagram is extended to two circles for ease of explanation of the calculations and the necessary change is shown in Figure 10.6. The left node computes the sum of the incoming inputs and this is indicated by the $+$ symbol on its right half. The activation function is then applied on the sum to obtain the quantity s and this node activity is captured in the right half of the second circle. The derivative of s is s' and it is displayed in the left half of the second circle. Since the derivative of any sum w.r.t. one of the sum argument is 1, this computation is captured in the left half of the first circle. Separation of integration and the differentiation steps shows the working of the node of an NN. Thus, the network is evaluated in two stages:

- *The Feed-forward Step.* Here, the information comes from the left-side and each unit evaluates the activation function on the right side and the derivative f' on the left side. The derivative is stored in the left side while the right-side component is transmitted further to the right side of the network.

- *The Backpropagation Step.* In this step, the whole network is run backward making use of the stored results in the process.

It can be easily seen from the structure of the NN that the backpropagation algorithm works for the unit node, nodes in series, and nodes in parallel. A more formal discussion of the backpropagation algorithm is taken up next.

10.5.1 Backpropagation Algorithm

We keep the discussion to a single input—single output NN structure for the classification problem. The feedforward step of a NN is straightforward. It is the calculation of the gradient as required for the neural network error correction that is more crucial to improvise the results. We assume that the activation functions of the NN are continuous and differentiable functions. Now, we begin with a network consisting of a single input and a single output. The input is denoted by x, the network function by F and its derivative by F'. The algorithm is given in the following:

ALGORITHM 10.5.1. **Backpropagation Algorithm**

1. Initialize the weights of the NN with random numbers.

2. **Feedforward Step**. This step consists of two substeps:

   (a) The input x is fed to the network and the activation function calculates the output after the node is traversed.

   (b) The derivative of the activation function is evaluated and stored.

3. **Backpropagation Step**. This step is carried out in the following:

   (a) The constant 1 is fed into the output unit and the network is run backward.

   (b) The incoming information (from the right side) to a node is added and the result is multiplied by the value stored by the left side of the B-diagram.

   (c) The result of the previous substep is transmitted to the left side of the node.

   (d) The result collected at the input unit is the derivative of the network function with respect to x.

4. Steps 2 and 3 are repeated until the difference between two consecutive iterations is less than a predecided small number.

We are not going into the mathematical proof that the backward pass calculates the gradient descent and refer the reader to Rojas (1996)[98] for comprehensive details.

Matt Mazur maintains a blog and he has a page dedicated to the numerical computational aspects of the backpropagation algorithm. It is to be noted that though the reader is likely to find thousands of blogs and articles on the backpropagation algorithm, a numerical illustration is rare to be found. The blog of Matt at `https://mattmazur.com/2015/03/17/a-step-by-step-backpropagation-example/` accomplishes that. The implementation in Matt's blog uses the Python software. However, we provide an R execution of the same. This is accomplished in the R file `../SRC/BackPropagation_01.R`. The entire forward pass and backpropagation steps are kept for a look in the file `../SRC/BackPropagation_02.R`. The object names are self-explanatory and we do not have to explain them any further. The reader must go through these two files to understand the computations associated with the neural networks.

In the concluding section of the chapter, we apply neural networks to survival data.

10.6   Neural Networks for Survival Analysis

We have seen the working of a neural network in previous sections. The technique has been developed with the intent of determining weights such that predicted values are as close as possible to the actuals. Note however that the purpose of determining the weights is not to identify the significant/causal covariates and this gives an impression that a neural network is a black box. This is even more so because the multiple layers mix up the impact of all the variables and once the mixing happens, the original impact of the variables looses traceability. In brief, NN as a methodology is concerned about prediction problems and not impact of specific covariates. Most of the applications in clinical trials studies relate to survival analysis. Ripley and Ripley (2007)[95] note the following attempts to apply the (classification) NN for survival data:

- Consider survival time up to a fixed time t as an indicator variable. Ignore all observations which were censored before time t and setup the classification NN.

- Bin the survival times into one of the k-time intervals and use the NN with multiple outputs.

- Create k separate NNs for the binning done in the previous step.

We will now see application of the classification NN for the PBC dataset. The structure of a few neural networks is visualized first. The `plotnet` function from the `NeuralNetTools` will be used to visualize the structure of three neural networks. All the networks contain two hidden layers. The number of neurons in the hidden layers are changed first, and then we look at a neural network with two neurons in the output layer.

```
> pdf("../Output/Neural_Networks.pdf")
> par(mfrow=c(3,1))
> plotnet(rep(0,17),struct=c(3,2,2,1))
> title("A Neural Network with Two Hidden Layers")
> plotnet(rep(0,39),struct=c(5,3,4,1))
> title("Another Neural Network with Two Hidden Layers")
> plotnet(rep(0,29),struct=c(6,3,1,2))
> title("A Neural Network with Two Output Neurons")
> dev.off()
```

Example 52 *Neural Network Analysis of PBC Dataset. We will continue to analyze the PBC dataset with the aid of neural networks. As mentioned earlier, we will bin the survival times and focus on the problem of whether the patient survived for more than 2000 days or not. As mentioned*

FIGURE 10.7: Neural Networks with Hidden Layers

earlier, when an observation is censored before the threshold, we will chuck out the observation. We will also remove all observations that have missing values among the covariates.

```
> pbc_nn <- pbc
> pbc_nn$tg2000 <- ifelse(pbc_nn$time>2000,"Long","Average")
> table(pbc_nn$tg2000)

Average    Long
   240      178
> sum(pbc_nn$time <= 2000 & pbc_nn$status!=2)
[1] 122
> pbc_nn <- pbc_nn[!(pbc_nn$time <= 2000 & pbc_nn$status!=2),]
> pbc_nn <- subset(pbc_nn,select = -c(id,time,status))
> pbc_nn$tg2000 <- as.factor(pbc_nn$tg2000)
> pbc_nn <- na.omit(pbc_nn)
> dim(pbc_nn)
[1] 199  18
> table(pbc_nn$tg2000)

Average    Long
```

77   122

We have 199 observations with no missing data. The patients who have survived for more than 2000 days are 122 in count while those failing before 2000 days number 77. The observations are labeled Long *and* Average *respectively. Our purpose is to fit a neural network for the dataset and then look at the accuracy. We will begin with 5 neurons in the hidden layer. We will step up the number of neurons to 10, 20, and finally 30. For the last choice, the convergence is not accomplished within 100 iterations and it is at iteration 150 that convergence is reached.*

```
> # creating the neural network
> pbc_NN_5 <- nnet(tg2000~.,data=pbc_nn,size=5)
# weights:  96
initial  value 131.890549
iter  10 value 119.468262
iter  20 value 119.105290
iter  30 value 118.831784
iter  40 value 107.012692
iter  50 value 103.858926
iter  60 value 103.616851
iter  70 value 100.810458
iter  80 value 99.624187
iter  90 value 96.242702
iter 100 value 95.686552
final  value 95.686552
stopped after 100 iterations
> pbc_predict_5 <- predict(pbc_NN_5,newdata=pbc_nn,type="class")
> sum(pbc_predict_5==pbc_nn$tg2000)
[1] 151
> table(pbc_predict_5,pbc_nn$tg2000)

pbc_predict_5 Average Long
      Average       71    42
      Long           6    80
> pbc_NN_10 <- nnet(tg2000~.,data=pbc_nn,size=10)
# weights:  191
initial  value 141.506716
iter  10 value 132.092496
final  value 132.092492
converged
> pbc_predict_10 <- predict(pbc_NN_10,newdata=pbc_nn,type="class")
> sum(pbc_predict_10==pbc_nn$tg2000)
[1] 122
> table(pbc_predict_10,pbc_nn$tg2000)
```

```
pbc_predict_10 Average Long
          Long      77  122
> pbc_NN_20 <- nnet(tg2000~.,data=pbc_nn,size=20)
# weights:  381
initial  value 190.917397
iter  10 value 125.472256
iter  20 value 120.250671
iter  30 value 117.840661
iter  40 value 113.011173
iter  50 value 108.388472
iter  60 value 107.299618
iter  70 value 106.207431
iter  80 value 103.191641
iter  90 value 101.131665
iter 100 value 100.540775
final  value 100.540775
stopped after 100 iterations
> pbc_predict_20 <- predict(pbc_NN_20,newdata=pbc_nn,type="class")
> sum(pbc_predict_20==pbc_nn$tg2000)
[1] 140
> table(pbc_predict_20,pbc_nn$tg2000)

pbc_predict_20 Average Long
       Average    51   33
       Long       26   89
> pbc_NN_30 <- nnet(tg2000~.,data=pbc_nn,size=30,maxit=200)
# weights:  571
initial  value 210.471785
iter  10 value 128.174270
iter  20 value 120.394136
iter  30 value 116.463374
iter  40 value 114.503229
iter  50 value 112.641819
iter  60 value 112.559003
iter  70 value 112.557340
iter  80 value 112.556195
iter  90 value 112.475330
iter 100 value 111.943209
iter 110 value 111.895171
iter 120 value 110.730599
iter 130 value 110.536630
iter 140 value 110.533530
iter 150 value 110.533392
iter 150 value 110.533392
iter 150 value 110.533392
```

```
final   value 110.533392
converged
> pbc_predict_30 <- predict(pbc_NN_30,newdata=pbc_nn,type="class")
> sum(pbc_predict_30==pbc_nn$tg2000)
[1] 137
> table(pbc_predict_30,pbc_nn$tg2000)
```

```
pbc_predict_30 Average Long
      Average      67    52
      Long         10    70
```

In the initial stage, we have predicted 151 observations out of 199 correctly with five neurons in the hidden layer. As the number of neurons is increased from 10 to 20 and then to 30, the accurately classified observations number respectively 122, 140, and 137. Recall that we begin the neural networks with random weights. Another run of the neural networks gets us the following output:

```
> sum(pbc_predict_5==pbc_nn$tg2000)
[1] 135
> sum(pbc_predict_10==pbc_nn$tg2000)
[1] 122
> sum(pbc_predict_20==pbc_nn$tg2000)
[1] 154
> sum(pbc_predict_30==pbc_nn$tg2000)
[1] 167
```

Multiple runs of the neural networks are required. Another option is to set the seeds at each run so that the fitted neural networks are reproducible. In the machine learning tradition, it is important for us to partition the dataset into train and test parts. The reader is recommended to try it out. Overall, we might say that convergence is the important criterion, and it does not matter how many neurons are in the network. The out-of-sample accuracy is also vital for deployment purposes.

□

Kvamme, et al. (2019)[68] adapt the Cox proportional hazards model for application of methods of neural networks. Kvamme and team also provide a Python implementation of their work and that appears to be most apt for dealing with survival data in NN.

10.7   Exercises

Exercise 10.1 *How many weights are required to be calculated for the PBC dataset if all the covariates are passed through the input layer and we place* *five neurons in the hidden layer? For simplicity, assume that we are modeling only the observed lifetime, setting aside the event indicator variable.* □

Exercise 10.2 *A crude way of modeling for survival data is to create two neural networks: the first neural network models the observed time as a numeric output and we run* nnet *with the option* linout = TRUE, *and a second neural network for the event indicator as a classification output. The predicted output of the two neural networks can then be combined for prediction problem. Carry out this analysis for the PBC dataset. Use two neurons in the hidden layers,* size = 2. □

Exercise 10.3 *Continue the previous exercise by increasing the number of neurons in the hidden layers to 5 and 10. Do you run into an error?* □

10.8 Más Lejos Temas

Use of Neural Networks has not seen adaptability and acceptance in survival data as much as the decision tree variant in survival tree. However, this is not an indication no attempts have been made to develop a version of NN for applicabity to survival data. Several attempts have gone into adaptation of NN for left truncated and right censored data. Ripley and Ripley (2007)[95] gives a number of references and additional related material. Further works can be found in Dybowski and Gant (2007)[38]. Cheng and Titterington (1994)[23] is an example of earlier instance of the NN application.

Kvamme, et al. (2019)[68] attempt prediction for time-to-event data using neural networks. Ranganath, et al. (2016)[92] develop deep survival networks.

Chapter 11

Complementary Machine Learning Techniques

Machine Learning (ML) offers a powerful narrative to the conventional statistical analysis of survival data. In the previous three chapters, we introduced just two families of ML techniques—trees and neural networks. ML encompasses a much wider family of techniques which have not been considered in this book for two reasons. The first is that it is not within the scope to cover the topics. The second reason is also that such techniques have not seen a proper extension to the survival data. In this section, we will outline a few topics which can be considered by the reader as topics of further interest.

One rich class of methods in the ML domain is the use of k-nearest neighborhoods, or simply k-NN. The idea here is quite simple and intuitive. To predict the y values, we look at the k-nearest neighbors of the observation by their $\mathbf{x}$ values. Nearest in what sense? If all components of the feature vector are numeric in nature, the nearness is best captured by the Euclidean distance. When we have mixture of numeric and categorical variables, we can employ the Manhattan distance. Using the concept of the distance, we find the k nearest neighbors of an observation. If we have a classification problem, the majority vote (class) of the k nearest neighbors labels is taken into account and that is returned as the prediction for the observation. In the case of the regression problem, the average of the y values of the nearest neighbors is taken as the prediction. Figure 11.1, obtained from Tattar (2018)[112], gives a depiction of the nearest neighbors and the corresponding class prediction.

The `bnnSurvival` package accomplishes the k-NN for survival data. The only parameter in the context of k-NN is the number of neighbors k. Of course, one might argue that the choice of the distance function, where two or more distance functions are possible, is also a parameter. Cross-validations help in determining the appropriate choice. While the k-NN is known to be useful for the classification problem, their usage generally does not yield good generalizations in regression problems. The k-NN needs to be used judiciously with survival data, and more so because of the censoring problems in survival data.

Multivariate survival data has not been handled in this text. We note that with multivariate survival data, the complexities are more pronounced. For instance, the bivariate survival function and the bivariate hazard rate are not easy to tackle. Hence, it becomes even more difficult to deal with multivariate

DOI: 10.1201/9781003306979-11

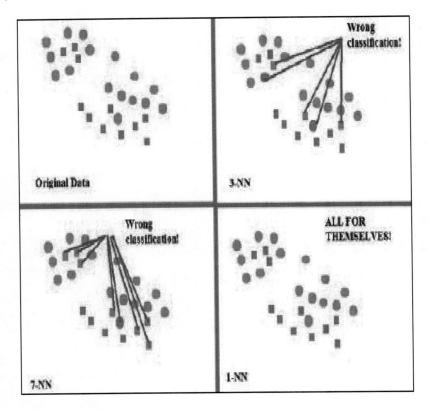

FIGURE 11.1: Illustrating k-NN

survival data in the machine learning domain. Besides, the frailty extension of the survival models does not have straightforward extension in the machine learning domain.

Deep Neural Networks have profound applications in face recognition, image classification, and video analysis. The extension of neural networks has seen lot of hectic activity in the past five to six years. The ease and cheap computational power has propelled research in this direction. Recurrent neural networks (RNN) and convolutional neural networks (CNN) have seen explosive growth and enabled the advancement of NN. Why do we have to bother with these developments? It is important more than ever before, and in light of the current and ongoing pandemic of Corona crisis. For instance, using the computed tomography scan, aka CT scan, is it possible for us to predict more accurately, on an average, how long an infected patient can live without lapsing into the stage where he needs oxygen support or antiviral like Remdeviser? If we can augment survival analysis with such impactful analysis, it will go a long way in planning the crisis which is impacting humanity in an unprecedented way. Not just identification, if the dose of the vials and other essential medicines can be determined with greater accuracy, a number of lives could be saved.

In deep NN, multiple layers of hidden neurons are used as against a single layer used in the previous chapter. It takes a lot of expertise and practice to fine tune NN toward more accurate and robust networks. Similarly, in CNN and RNN analysis, networks help in sequence analysis. In the pandemic context, it becomes important to be able to predict how many patients are likely to face the black fungus problem after recovering from the virus attack. Thus, we need to be prepared in the usage of these ML techniques too.

Finally, the ensemble techniques seen in this part combine only trees. Heterogeneous ensemble is desperately required. Predictions are not easily obtainable in GBM. We need definite progress in these directions too. Machine learning and traditional survival analysis should go hand-in-hand. We hope the reader will be able to contribute in this broader area.

Bibliography

[1] Odd Aalen. A model for nonparametric regression analysis of counting processes. In *Mathematical statistics and probability theory*, pages 1–25. Springer, 1980.

[2] Odd Aalen, Ornulf Borgan, and Hakon Gjessing. *Survival and event history analysis: a process point of view*. Springer Science & Business Media, 2008.

[3] Odd O Aalen and Søren Johansen. An empirical transition matrix for non-homogeneous markov chains based on censored observations. *Scandinavian Journal of Statistics*, pages 141–150, 1978.

[4] OO Aalen. *Statistical inference for a family of counting process*. PhD thesis, University of California, Berkeley, 1975.

[5] Htrotugu Akaike. Maximum likelihood identification of gaussian autoregressive moving average models. *Biometrika*, 60(2):255–265, 1973.

[6] Michael G Akritas. Bootstrapping the kaplan—meier estimator. *Journal of the American Statistical Association*, 81(396):1032–1038, 1986.

[7] Per K Andersen, Ornulf Borgan, Richard D Gill, and Niels Keiding. *Statistical models based on counting processes*. Springer Science & Business Media, 1993.

[8] Per Kragh Andersen, Mette Gerster Hansen, and John P Klein. Regression analysis of restricted mean survival time based on pseudo-observations. *Lifetime data analysis*, 10(4):335–350, 2004.

[9] Per Kragh Andersen, John P Klein, and Susanne Rosthøj. Generalised linear models for correlated pseudo-observations, with applications to multi-state models. *Biometrika*, 90(1):15–27, 2003.

[10] Narayanaswamy Balakrishnan and Chin Diew Lai. *Continuous bivariate distributions*. Springer Science & Business Media, 2009.

[11] Jan Beyersmann, Arthur Allignol, and Martin Schumacher. *Competing risks and multistate models with R*. Springer Science & Business Media, 2011.

[12] Peter J Bickel, Chris AJ Klaassen, Peter J Bickel, Ya'acov Ritov, J Klaassen, Jon A Wellner, and YA'Acov Ritov. *Efficient and adaptive estimation for semiparametric models*, volume 4. Johns Hopkins University Press Baltimore, 1993.

[13] Ørnulf Borgan. Maximum likelihood estimation in parametric counting process models, with applications to censored failure time data. *Scandinavian Journal of Statistics*, pages 1–16, 1984.

[14] Imad Bou-Hamad, Denis Larocque, and Hatem Ben-Ameur. A review of survival trees. *Statistics surveys*, 5:44–71, 2011.

[15] George EP Box and David R Cox. An analysis of transformations. *Journal of the Royal Statistical Society: Series B (Methodological)*, 26(2):211–243, 1964.

[16] Leo Breiman. Bagging predictors. *Machine learning*, 24(2):123–140, 1996.

[17] Leo Breiman. Random forests. *Machine learning*, 45(1):5–32, 2001.

[18] Leo Breiman. Statistical modeling: The two cultures (with comments and a rejoinder by the author). *Statistical science*, 16(3):199–231, 2001.

[19] Leo Breiman, Jerome H Friedman, Richard A Olshen, and Charles J Stone. *Classification and regression trees*. Routledge, 1984.

[20] Göran Broström. *Event history analysis with R*. CRC Press, 2018.

[21] Jonathan Buckley and Ian James. Linear regression with censored data. *Biometrika*, 66(3):429–436, 1979.

[22] George Casella and Roger L Berger. *Statistical inference*. Cengage Learning, 2002.

[23] Bing Cheng and D Michael Titterington. Neural networks: A review from a statistical perspective. *Statistical science*, pages 2–30, 1994.

[24] A Ciampi, RS Bush, M Gospodarowicz, and JE Till. An approach to classifying prognostic factors related to survival experience for non-hodgkin's lymphoma patients: Based on a series of 982 patients: 1967–1975. *Cancer*, 47(3):621–627, 1981.

[25] Gerda Claeskens and Nils Lid Hjort. The focused information criterion. *Journal of the American Statistical Association*, 98(464):900–916, 2003.

[26] Gerda Claeskens and Nils Lid Hjort. Model selection and model averaging. *Cambridge Books*, 2008.

[27] R Dennis Cook and Sanford Weisberg. *Residuals and influence in regression*. New York: Chapman and Hall, 1982.

[28] David R Cox. Regression models and life-tables. *Journal of the Royal Statistical Society: Series B (Methodological)*, 34(2):187–202, 1972.

[29] David R Cox. Partial likelihood. *Biometrika*, 62(2):269–276, 1975.

[30] David Roxbee Cox and David Oakes. *Analysis of survival data*. Chapman and Hall/CRC, 1984.

[31] D.R. Cox. Some simple approximate tests for poisson variates. *Biometrika*, 40(3/4):354–360, 1953.

[32] DR Cox and EJ Snell. Applied statistics. 1984.

[33] Peter Dalgaard. *Introductory statistics with R, 2e*. Springer, 2008.

[34] Herbert Aron David and Melvin L Moeschberger. *The Theory of Competing Risks: HA David, ML Moeschberger*. C. Griffin, 1978.

[35] Roger B Davis and James R Anderson. Exponential survival trees. *Statistics in medicine*, 8(8):947–961, 1989.

[36] Arthur P Dempster, Nan M Laird, and Donald B Rubin. Maximum likelihood from incomplete data via the em algorithm. *Journal of the Royal Statistical Society: Series B (Methodological)*, 39(1):1–22, 1977.

[37] Luc Duchateau and Paul Janssen. *The frailty model*. Springer, 2008.

[38] Richard Dybowski and Vanya Gant. *Clinical applications of artificial neural networks*. Cambridge University Press, 2007.

[39] B. Efron. Bootstrap Methods: Another Look at the Jackknife. *The Annals of Statistics*, 7(1):1 – 26.

[40] Bradley Efron. Censored data and the bootstrap. *Journal of the American Statistical Association*, 76(374):312–319, 1981.

[41] Bradley Efron and Trevor Hastie. *Computer age statistical inference*, volume 5. Cambridge University Press, 2016.

[42] Regina C Elandt-Johnson and Norman L Johnson. *Survival models and data analysis*, volume 110. John Wiley & Sons, 1980.

[43] Jianqing Fan and Runze Li. Variable selection via nonconcave penalized likelihood and its oracle properties. *Journal of the American statistical Association*, 96(456):1348–1360, 2001.

[44] Thomas R Fleming and David P Harrington. *Counting processes and survival analysis*, volume 169. John Wiley & Sons, 1991.

[45] Yoav Freund, Robert E Schapire, et al. Experiments with a new boosting algorithm. In *icml*, volume 96, pages 148–156. Citeseer, 1996.

[46] Jerome H Friedman. Greedy function approximation: a gradient boosting machine. *Annals of statistics*, pages 1189–1232, 2001.

[47] Jelle J Goeman. L1 penalized estimation in the cox proportional hazards model. *Biometrical journal*, 52(1):70–84, 2010.

[48] Louis Gordon and Richard A Olshen. Tree-structured survival analysis. *Cancer treatment reports*, 69(10):1065–1069, 1985.

[49] Anders Gorst-Rasmussen and Thomas H Scheike. Coordinate descent methods for the penalized semiparametric additive hazards model. *Journal of Statistical Software*, 47(1):1–17, 2012.

[50] Rameshwar D Gupta and Debasis Kundu. A new class of weighted exponential distributions. *Statistics*, 43(6):621–634, 2009.

[51] Trevor Hastie, Robert Tibshirani, and Jerome Friedman. *The Elements of Statistical Learning, 2e*. Springer Series in Statistics. Springer New York Inc., 2009.

[52] Simon S Haykin. *Neural networks and learning machines*. Prentice Hall, 2009.

[53] Nils Lid Hjort. Focused information criteria for the linear hazard regression model. In *Statistical models and methods for biomedical and technical systems*, pages 487–502. Springer, 2008.

[54] Nils Lid Hjort and Gerda Claeskens. Focused information criteria and model averaging for the cox hazard regression model. *Journal of the American Statistical Association*, 101(476):1449–1464, 2006.

[55] Torsten Hothorn, Berthold Lausen, Axel Benner, and Martin Radespiel-Tröger. Bagging survival trees. *Statistics in medicine*, 23(1):77–91, 2004.

[56] Aparna V Huzurbazar. *Flowgraph models for multistate time-to-event data*, volume 439. John Wiley & Sons, 2005.

[57] Hemant Ishwaran, Udaya B Kogalur, Eugene H Blackstone, and Michael S Lauer. Random survival forests. *The annals of applied statistics*, 2(3):841–860, 2008.

[58] L James. General exchangeable bootstraps for the aalen–johansen estimator. *Unpublished manuscript*, 1998.

[59] Norman L Johnson, Samuel Kotz, and Narayanaswamy Balakrishnan. *Continuous univariate distributions, volume 2*. John wiley & sons, 1995.

[60] John D Kalbfleisch and Ross L Prentice. *The statistical analysis of failure time data, 2e*. John Wiley & Sons, 2002.

[61] Edward L Kaplan and Paul Meier. Nonparametric estimation from incomplete observations. *Journal of the American statistical association*, 53(282):457–481, 1958.

[62] Alan F Karr. *Point Processes and Their Statistical Inference*. Marcel Dekker, 1986.

[63] John P Klein and Per Kragh Andersen. Regression modeling of competing risks data based on pseudovalues of the cumulative incidence function. *Biometrics*, 61(1):223–229, 2005.

[64] John P Klein and Melvin L Moeschberger. *Survival analysis: techniques for censored and truncated data*, volume 1230. Springer, 2003.

[65] Michael R Kosorok. *Introduction to empirical processes and semiparametric inference*. Springer, 2008.

[66] H Koul, Vyaghreswarudu Susarla, and John Van Ryzin. Regression analysis with randomly right-censored data. *The Annals of statistics*, pages 1276–1288, 1981.

[67] Ludmila I Kuncheva. *Combining pattern classifiers: methods and algorithms, 2e*. John Wiley & Sons, 2014.

[68] Håvard Kvamme, Ørnulf Borgan, and Ida Scheel. Time-to-event prediction with neural networks and cox regression. *arXiv preprint arXiv:1907.00825*, 2019.

[69] Jerald F Lawless. *Statistical models and methods for lifetime data, 2e*, volume 362. John Wiley & Sons, 2002.

[70] Michael LeBlanc and John Crowley. Relative risk trees for censored survival data. *Biometrics*, pages 411–425, 1992.

[71] Elisa T Lee and John Wang. *Statistical methods for survival data analysis, 3e*, volume 476. John Wiley & Sons, 2003.

[72] Dan Yu Lin and Zhiliang Ying. Semiparametric analysis of the additive risk model. *Biometrika*, 81(1):61–71, 1994.

[73] Roderick JA Little and Donald B Rubin. *Statistical analysis with missing data, 2e*, volume 793. John Wiley & Sons, 2002.

[74] Albert W Marshall and Ingram Olkin. *Life distributions*, volume 13. Springer, 2007.

[75] Albery W Marshall and Ingram Olkin. A generalized bivariate exponential distribution. Technical report, BOEING SCIENTIFIC RESEARCH LABS SEATTLE WA MATHEMATICS RESEARCH LAB, 1966.

[76] Torben Martinussen and Thomas H Scheike. *Dynamic regression models for survival data.* Springer Science & Business Media, 2006.

[77] Torben Martinussen and Thomas H Scheike. Aalen additive hazards change-point model. *Biometrika*, 94(4):861–872, 2007.

[78] E Marubini, A Morabito, and MG Valsecchi. Prognostic factors and risk groups: some results given by using an algorithm suitable for censored survival data. *Statistics in medicine*, 2(2):295–303, 1983.

[79] Ian W McKeague and Peter D Sasieni. A partly parametric additive risk model. *Biometrika*, 81(3):501–514, 1994.

[80] Geoffrey J McLachlan and Thriyambakam Krishnan. *The EM algorithm and extensions*, volume 382. John Wiley & Sons, 2008.

[81] Rupert G Miller. Least squares regression with censored data. *Biometrika*, 63(3):449–464, 1976.

[82] Rupert G Miller. *Survival Analysis.* John Wiley & Sons, 1981.

[83] John F Monahan. *Numerical methods of statistics.* Cambridge University Press, 2011.

[84] Douglas C Montgomery, Elizabeth A Peck, and G Geoffrey Vining. *Introduction to linear regression analysis.* John Wiley & Sons, 2021.

[85] Dirk Foster Moore. *Applied survival analysis using R.* Springer, 2016.

[86] James N Morgan and John A Sonquist. Problems in the analysis of survey data, and a proposal. *Journal of the American statistical association*, 58(302):415–434, 1963.

[87] Saralees Nadarajah and Samuel Kotz. Reliability for some bivariate exponential distributions. *Mathematical Problems in Engineering*, 2006, 2006.

[88] Wayne B Nelson. *Applied life data analysis*, volume 521. John Wiley & Sons, 1982.

[89] John O'Quigley. *Proportional hazards regression*, volume 542. Springer, 2008.

[90] Richard Peto and Julian Peto. Asymptotically efficient rank invariant test procedures. *Journal of the Royal Statistical Society: Series A (General)*, 135(2):185–198, 1972.

[91] J Ross Quinlan. *C4. 5: programs for machine learning.* Elsevier, 1993.

[92] Rajesh Ranganath, Adler Perotte, Noémie Elhadad, and David Blei. Deep survival analysis. In *Machine Learning for Healthcare Conference*, pages 101–114. PMLR, 2016.

[93] Greg Ridgeway. Generalized boosted models: A guide to the gbm package. *Update*, 1(1):2007, 2007.

[94] Brian D Ripley. Statistical aspects of neural networks. *Networks and chaos-statistical and probabalistic aspects*, pages 40–123, 1993.

[95] Brian D Ripley and Ruth M Ripley. Neural networks as statistical methods in survival analysis. *Clinical applications of artificial neural networks*, 237:255, 2007.

[96] James M Robins and Andrea Rotnitzky. Recovery of information and adjustment for dependent censoring using surrogate markers. In *AIDS epidemiology*, pages 297–331. Springer, 1992.

[97] Vijay K Rohatgi and AK Md Ehsanes Saleh. *An introduction to probability and statistics*. John Wiley & Sons, 2015.

[98] Raúl Rojas. *Neural networks: a systematic introduction*. Springer Science & Business Media, 1996.

[99] Joseph L Schafer. *Analysis of incomplete multivariate data*. CRC press, 1997.

[100] Robert E Schapire and Yoav Freund. *Boosting: Foundations and Algorithms. Adaptive computation and machine learning*. Mit Press London, 2012.

[101] Gideon Schwarz. Estimating the dimension of a model. *The annals of statistics*, pages 461–464, 1978.

[102] Guido Schwarzer, Werner Vach, and Martin Schumacher. On the misuses of artificial neural networks for prognostic and diagnostic classification in oncology. *Statistics in medicine*, 19(4):541–561, 2000.

[103] GAF Seber and CJ Wild. *Nonlinear regression*. Wiley, New York, 1989.

[104] Shai Shalev-Shwartz and Shai Ben-David. *Understanding machine learning: From theory to algorithms*. Cambridge university press, 2014.

[105] Linda D Sharples, Christopher H Jackson, Jayan Parameshwar, John Wallwork, and Stephen R Large. Diagnostic accuracy of coronary angiography and risk factors for post–heart-transplant cardiac allograft vasculopathy. *Transplantation*, 76(4):679–682, 2003.

[106] Galen R Shorack and Jon A Wellner. *Empirical processes with applications to statistics*. John Wiley & Sons, 1986.

[107] Peter J Smith. *Analysis of failure and survival data*. CRC Press, 2002.

[108] Winfried Stute. Kaplan–meier integrals. *Handbook of Statistics*, 23:87–104, 2003.

[109] Pang-Ning Tan, Michael Steinbach, and Vipin Kumar. *Introduction to data mining*. Pearson, 2005.

[110] Prabhanjan Narayanachar Tattar. Bootstrap test of hypothesis for the multi-state models in survival analysis. *Communications in Statistics-Theory and Methods*, 45(5):1270–1277, 2016.

[111] Prabhanjan Narayanachar Tattar. *Statistical Application Development with R and Python, 2e*. Packt Publishing Ltd, 2017.

[112] Prabhanjan Narayanachar Tattar. *Hands-On Ensemble Learning with R*. Packt Publishing Ltd, 2018.

[113] Prabhanjan Narayanachar Tattar, Suresh Ramaiah, and B G Manjunath. *A Course in Statistics with R*. John Wiley & Sons, 2016.

[114] Prabhanjan Narayanachar Tattar and H Jalikop Vaman. Testing transition probability matrix of a multi-state model with censored data. *Lifetime data analysis*, 14(2):216–230, 2008.

[115] Prabhanjan Narayanachar Tattar and H Jalikop Vaman. The k-sample problem in a multi-state model and testing transition probability matrices. *Lifetime data analysis*, 20(3):387–403, 2014.

[116] Terry M Therneau and Patricia M Grambsch. *Modeling survival data: extending the Cox model*. Springer, 2000.

[117] Terry M Therneau, Patricia M Grambsch, and Thomas R Fleming. Martingale-based residuals for survival models. *Biometrika*, 77(1):147–160, 1990.

[118] Robert Tibshirani. Regression shrinkage and selection via the lasso. *Journal of the Royal Statistical Society: Series B (Methodological)*, 58(1):267–288, 1996.

[119] Frans Willekens. *Multistate analysis of life histories with R*. Springer, 2014.

[120] Heping Zhang and Burton H Singer. *Recursive partitioning and applications*. Springer Science & Business Media, 2010.

[121] Hongwei Zhao and Anastasios A Tsiatis. Testing equality of survival functions of quality-adjusted lifetime. *Biometrics*, 57(3):861–867, 2001.

Index

For Product Safety Concerns and Information please contact our EU
representative GPSR@taylorandfrancis.com Taylor & Francis Verlag GmbH,
Kaufingerstraße 24, 80331 München, Germany

Printed and bound by CPI Group (UK) Ltd, Croydon, CR0 4YY
06/05/2025
01861444-0001